Communications in Computer and Information Science 1352

More information about this series at http://www.springer.com/series/7899

Kun He · Cheng Zhong ·
Zhiping Cai · Yitong Yin (Eds.)

Theoretical Computer Science

38th National Conference, NCTCS 2020
Nanning, China, November 13–15, 2020
Revised Selected Papers

Editors
Kun He 🄳
Huazhong University of Science
and Technology
Wuhan, China

Cheng Zhong 🄳
Guangxi University
Nanning, China

Zhiping Cai 🄳
National University of Defense Technology
Changsha, China

Yitong Yin 🄳
Nanjing University
Nanjing, China

ISSN 1865-0929   ISSN 1865-0937   (electronic)
Communications in Computer and Information Science
ISBN 978-981-16-1876-5   ISBN 978-981-16-1877-2   (eBook)
https://doi.org/10.1007/978-981-16-1877-2

This Springer imprint is published by the registered company Springer Nature Singapore Pte Ltd.
The registered company address is: 152 Beach Road, #21-01/04 Gateway East, Singapore 189721, Singapore

Preface

The National Conference of Theoretical Computer Science (NCTCS) is the main academic activity in the area of theoretical computer science in China. Up until 2020, NCTCS has been successfully held 37 times in over 20 cities. It provides a platform for researchers in the area of theoretical computer science or related areas for academic exchanges and possible cooperation.

This volume contains the papers presented at NCTCS 2020: The 38th National Conference of Theoretical Computer Science held during November 13–15, 2020, in Nanning, China. Sponsored by the China Computer Forum (CCF), NCTCS 2020 was hosted by the CCF Theoretical Computer Science Committee and Guangxi University.

NCTCS 2020 received 28 submissions (in English) in the areas of algorithms and complexity, theory or algorithm aspects for data science and deep learning, network communication, and security. Each of the 28 submissions was reviewed by at least three Program Committee members. The committee decided to accept 13 papers that are included in this Springer Communications in Computer and Information Science (CCIS) publication.

NCTCS 2020 invited prestigious researchers carrying out a wide range of academic activities in the field of theoretical computer science to give a keynote speech, and introduce their recent advanced research results. We had six invited plenary speakers at NCTCS 2020: Huimin LIN (Institute of Software, Chinese Academy of Sciences, China), Haimin WANG (National Defense University of Science and Technology, China), Wenfei YU (University of Edinburgh, UK), Xiaomin LI (Peking University, China), Zhiwei XU (Institute of Computing Technology, Chinese Academy of Sciences, China), and Lian LI (Hefei University of Technology, China). We express our sincere thanks to them for their contributions to the conference and proceedings.

The proceedings editors wish to thank the dedicated Program Committee members and external reviewers for their hard work in reviewing and selecting papers. We also thank Springer for their trust and for publishing the proceedings of NCTCS 2020.

December 2020

Kun He
Cheng Zhong
Zhiping Cai
Yitong Yin

Organization

General Chairs

Xiaoming Sun   Institute of Computing Technology, Chinese Academy
                      of Sciences, China
Jinzhao Wu   Guangxi University, China

Program Chairs

Kun He   Huazhong University of Science and Technology,
                      China
Cheng Zhong   Guangxi University, China
Zhiping Cai   National University of Defense Technology, China
Yitong Yin   Nanjing University, China

Organization Chairs

Cheng Zhong   Guangxi University, China
Zhenrong Zhang   Guangxi University, China
Jialin Zhang   Institute of Computing Technology, Chinese Academy
                      of Sciences, China

Program Committee

Xiaomin Sun   Institute of Computing Technology, Chinese Academy
                      of Sciences, China
Xiaohui Bei   Nanyang Technological University, China
Kerong Ben   Naval University of Engineering, China
Yixin Cao   Hong Kong Polytechnic University, China
Yongzhi Cao   Peking University, China
Juan Chen   National University of Defense Technology, China
Xiaoyun Chen   Lanzhou University, China
Heng Guo   University of Edinburgh, UK
Kun He   Huazhong University of Science and Technology,
                      China
Zhenyan Ji   Beijing Jiaotong University, China
Haitao Jiang   Shandong University, China
Chu-Min Li   University of Picardie Jules Verne, France
Dongkui Li   Baotou Teachers College, China
Jian Li   Tsinghua University, China
Lvzhou Li   Zhongshan University, China
Minming Li   City University of Hong Kong, China

Zhanshan Li	Jilin University, China
Peiqiang Liu	Shandong Technology and Business University, China
Qiang Liu	National University of Defense Technology, China
Tian Liu	Peking University, China
Xiaoguang Liu	Nankai University, China
Xinwang Liu	National University of Defense Technology, China
Zhendong Liu	Shandong Jianzhu University, China
Shuai Lu	Jilin University, China
Rui Mao	Shenzhen University, China
Xinjun Mao	National University of Defense Technology, China
Dantong Ouyang	Jilin University, China
Jianmin Pang	Information Engineering University, China
Zhiyong Peng	Wuhan University, China
Zhengwei Qi	Shanghai Jiao Tong University, China
Junyan Qian	Guilin University of Electronic Technology, China
Jiaohua Qin	Hunan City University, China
Kaile Su	Griffith University, Australia
Xiaoming Sun	Chinese Academy of Sciences, China
Chang Tang	China University of Geosciences, China
Gang Wang	Nankai University, China
Jianxin Wang	Central South University, China
Liwei Wang	Peking University, China
Zihe Wang	Shanghai University of Finance and Economics, China
Jigang Wu	Guangdong University of Technology, China
Zhengjun Xi	Shaanxi Normal University, China
Mingji Xia	Chinese Academy of Sciences, China
Meihua Xiao	East China Jiaotong University, China
Mingyu Xiao	University of Electronic Science and Technology of China, China
Jinyun Xue	Jiangxi Normal University, China
Yan Yang	Southwest Jiaotong University, China
Jianping Yin	National University of Defense Technology, China
Yitong Yin	Nanjing University, China
Mengting Yuan	Wuhan University, China
Defu Zhang	Xiamen University, China
Jialin Zhang	Chinese Academy of Sciences, China
Jianming Zhang	Changsha University of Science and Technology, China
Peng Zhang	Shandong University, China
Yong Zhang	Shenzhen Institutes of Advanced Technology, Chinese Academy of Sciences, China
Zhao Zhang	Zhejiang Normal University, China
Yujun Zheng	Zhejiang University of Technology, China
En Zhu	National University of Defense Technology, China

Contents

Network Communication and Security

Algorithm and Complexity

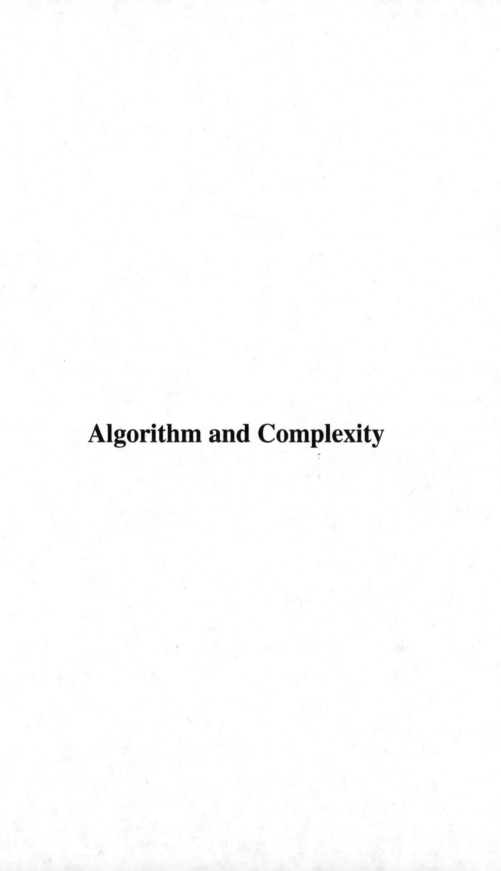

Accelerating Predicate Abstraction by Minimum Unsatisfiable Cores Extraction

Jianmin Zhang$^{(\boxtimes)}$, Tiejun Li, and Kefan Ma

School of Computer, National University of Defense Technology,
Changsha 410073, China
jmzhang@nudt.edu.cn

Abstract. With the growing scale and complexity of software and hardware designs, model checking generally results in the combination explosion of state space. Predicate abstraction is an important technique to solve this problem. The number of refinement iterations will be reduced by extracting the unsatisfiable cores. The smaller unsatisfiable cores are, the more false counterexamples are eliminated. Therefore, a fast algorithm of deriving the minimum unsatisfiable cores is employed in the formal verification tool of hardware. The two optimal algorithms of computing minimum unsatisfiable cores are compared on the instruction Cache unit of a microprocessor. The experimental results showed that the greedy-generic algorithm outperforms the branch-bound algorithm. Furthermore we analyzed that the unsatisfiable cores plays an important role in predicate abstraction, and it can improve the efficiency of model checking.

Keywords: Formal verification · Model checking · Predicate abstraction · Boolean satisfiability · Minimum unsatisfiable core

1 Introduction

Many real-world problems, arising in software verification, electronic design, equivalence checking, property verification and Auto Test Pattern Generation (ATPG), can be formulated as constraint satisfaction problems, which can be translated into Boolean formulas in conjunctive normal form (CNF). Modern Boolean satisfiability (SAT) solvers, such as Chaff [1] and MiniSAT [2], which implement enhanced versions of the Davis-Putnam-Logemann-Loveland (DPLL) backtrack-search algorithm, are usually able to determine whether a large formula is satisfiable or not. When a formula is unsatisfiable, it is often required to find an unsatisfiable core, that is, a small unsatisfiable subset of the original formula. Localizing a small unsatisfiable core is necessary to determine the underlying reasons for the failure.

Supported by the National Natural Science Foundation of China under grant No. 62072464 and U19A2062, and the National Laboratory of Parallel and Distributed Processing Open Fund under grant No. WDZC20205500116.

K. He et al. (Eds.): NCTCS 2020, CCIS 1352, pp. 3–13, 2021.
https://doi.org/10.1007/978-981-16-1877-2_1

Explaining the causes of unsatisfiability of Boolean formulas is an essential requirement in various fields, such as electronic design automation and formal verification of software and hardware. A typical paradigm is the predicate abstraction and refinement of model checking. Since the function of VLSI circuits is more and more complicated, the number of gates in the chips is getting larger. Then during the formal verification on the chips, the state space may result to a "combinational explosion" problem. Therefore, in both academic and industry fields, the abstract and refinement technology is generally used to map the original RTL design to an abstract model, and then performs model checking on the abstract model. In recent past, the predicate abstract method [3] is one of the current main abstract technologies, and has been addressed rather frequently in the last few years, owing to its increasing importance in numerous fields.

The predicate abstraction technology of hardware is to map the original state space of a Register Transfer Level (RTL) design or a gate-level design to a smaller state space, and then check whether the functional assertions are violated or not. If the properties are violated, a counterexample will be generated. Sometimes the false counterexamples are caused by over abstraction, and then the abstract model should be refined. The simulation process of the counterexample on the concrete model can be converted into a Boolean formula. If this formula is unsatisfiable, it indicates that a false counterexample has occurred. In the refinement process, these false counterexamples will be removed from the abstract model through unsatisfiable cores; and then on the refined abstract model, it will continue to loop until the properties are satisfied or produce the real counterexamples. In general, the size of the unsatisfiable cores determines the number of refining cycles, that is to say, it affects the efficiency of the entire verification process.

The unsatisfiable subformulas solver plays a very important role in the predicate abstraction and refinement of model checking for VLSI circuits. In general, the smaller unsatisfiable cores are, the more false counterexamples are eliminated, and the fewer number of refinement iterations are. Then the total runtime of formal verification of the designs will be reduced. Therefore, a fast algorithm of deriving the minimum unsatisfiable cores is employed in the property checking tool of hardware. A minimum unsatisfiable core has the smallest cardinality of all unsatisfiable subsets of a formula, and its size is generally much smaller than the size of minimal unsatisfiable ones. We have compared two optimal algorithms of computing minimum unsatisfiable cores, respectively called the branch-and-bound algorithm [4] and the greedy-genetic algorithm [5], on the instruction Cache unit of a 32-bit microprocessor PicoJava-II from Sun. The evaluation results show that the greedy-genetic algorithm strongly outperforms the branch-and-bound algorithm on runtime. It is also shown that the minimum unsatisfiable cores can help the model checking tool to accelerate the predicate abstraction and refinement process.

The paper is organized as follows. The next section gives the basic definitions and notations of unsatisfiable cores used throughout the paper. Section 3 surveys the related work on computing unsatisfiable cores. Section 4 introduces the architecture and predicate abstraction of PicoJava microprocessor. Section 5 shows and analyzes experimental results of two algorithm of extracting unsatisfiable cores on the instruction Cache unit. Finally, Sect. 6 concludes this paper.

2   Preliminaries

Resolution is a proof system for CNF (Conjunctive Normal Form) formulas with the following rule:

$$\frac{(A \lor x)(B \lor \neg x)}{(A \lor B)}, \tag{1}$$

where A, B are disjunctions of literals. The clauses $(A \lor x)$ and $(B \lor x)$ are the resolving clauses, and $(A \lor B)$ is the resolvent. The resolvent of the clauses (x) and $(\neg x)$ is the empty clause $(\bot)$. Each application of the resolution rule is called a resolution step. A sequence of resolution steps is that each one uses the result of the previous step or the clauses of the original formula as the resolving clauses of the current step.

Definition 1 (Boolean Satisfiability). *Given a CNF formula $\varphi(V)$, where V is the set of variables, and a Boolean function $F(V) : \{0,1\}^n \to \{0,1\}$, the Boolean satisfiability problem consists of identifying a set of assignments V' to the variables, such that $F(V') = 1$, or proving that no such assignment exists.*

Lemma 1. *A CNF formula φ is unsatisfiable iff there exists a finite sequence of resolution steps ending with the empty clause.*

It is well-known that a Boolean formula in CNF is unsatisfiable if it is possible to generate an empty clause by resolution sequence from the original clauses. The set of original clauses involved in the derivation of the empty clause is referred to as the unsatisfiable core.

Definition 2 (Unsatisfiable Core). *Given a formula φ, ϕ is an unsatisfiable core for φ iff ϕ is an unsatisfiable formula and $\phi \subseteq \varphi$.*

Observe that an unsatisfiable core can be defined as any subset, which is infeasible, of the original formula. Consequently, there may exist many different unsatisfiable cores, with different number of clauses, for the same problem instance, such that some of these cores are subsets of others.

Definition 3 (Minimal Unsatisfiable Core). *Given an unsatisfiable core ϕ for a formula φ, ϕ is a minimal unsatisfiable core if and only if removing any clause $\omega \in \phi$ from ϕ implies that $\phi - \{\omega\}$ is satisfiable.*

For Boolean formulas in CNF, an unsatisfiable core is minimal if it becomes satisfiable whenever any of its clauses is removed. According to the definition, a minimal unsatisfiable core has two features: one is unsatisfiable, the other is irreducible, in other words, all of its proper subsets are satisfiable.

Definition 4 *(Minimum Unsatisfiable Core). Consider the set of all unsat-isfiable cores for a formula* $\varphi : \{\phi_1, \ldots, \phi_j\}$. *Then,* $\phi_k \in \{\phi_1, \ldots, \phi_j\}$ *is a minimum unsatisfiable core iff* $\forall \; \phi_i \in \{\phi_1, \ldots, \phi_j\}$, $1 \leq i \leq j : |\phi_k| \leq |\phi_i|$.

That is to say, a minimum unsatisfiable core has the smallest cardinality of all unsatisfiable subsets of a formula. From the above definition, one may conclude that any unsatisfiable formula has at least one minimum unsatisfiable core. We should note that in many cases the size of a minimal unsatisfiable core may be much larger than the size of a minimum unsatisfiable core, because the minimal unsatisfiable cores probably contain many redundant clauses which cannot be found by simple resolution rules. Whereas, the existing works have very little attention on minimum unsatisfiable cores extraction and its applications.

3   Algorithms of Extracting Unsatisfiable Cores

There have been many different contributions to research on unsatisfiable cores extraction in the last few years, owing to the increasing importance in numerous practical applications. An algorithm [6] to derive a minimum unsatisfiable core enumerates all possible subsets of a formula and checks which are unsatisfiable but all of whose subsets are satisfiable. In this way, it can find the solution exactly, but the technique is unlikely to scale well. Some research works, based on a relationship between maximal satisfiability and minimal unsatisfiability, have developed some sound techniques for finding a minimum unsatisfiable core [4,5], or all minimal unsatisfiable cores [7].

CoreTimmer [8] iterates over each internal node that consumes a large number of clauses and attempts to prove them without these clauses. In [9], the authors presented the algorithms which tracks minimal unsatisfiable cores according to the trace of a failed local search run for consistency checking. Two new resolution-based algorithms are proposed in [10]. These algorithms are used to compute a minimal unsatisfiable core or, if time-out encountered, a small non-minimal unsatisfiable core. Based on these algorithms, seven improvements are proposed in [11], and the experiments have shown the reduction of 55% in run time and 73% in the size of the resulting subformula.

In [12], the authors also proposed two algorithms, one is to optimize the number of calls to a SAT solver, the other is to employ a new technique named recursive model rotation. An improvement to model rotation called eager rotation [13] is integrated in resolution-based minimal unsatisfiable subformulas algorithm. Anton et al. [14] have proposed some techniques to trim CNF formulas using clausal proofs. A new algorithm [15] is to exploit the minimal unsatisfiable subformulas (mus) and minimal correction sets connection in order to compute a single mus and to incrementally compute all muses. An algorithm [16] is proposed to improve over previous methods for finding multiple muses by computing its muses incrementally. The authors [17] aimed to explore the parallelization of partial MUS enumeration. The evaluation results show that the full parallelization of the entire enumeration algorithm scales well.

A technique [18] implements a model rotation paradigm that allows the set of minimal correction subsets to be computed in a heuristically efficient way. The authors [19] proposed a new algorithm for extracting minimal unsatisfiable cores and correction sets simultaneously. Liu et al. [20] intoduced an algorithm for extracting all muses for formulas in the field of propositional logic and the function-free and equality-free fragment of first-order logic. Luo et al. [21] proposed a method for accelerating the enumeration of muses based on inconsistency graph partitioning. A novel algorithm [22] is presented for computing the union of the clauses included in some muses, by developing a refined recursive enumeration of muses based on powerful pruning techniques.

Jaroslav et al. [23] firstly approximated mus counting procedure called AMU-SIC, combining the technique of universal hashing with advances in QBF solvers along with a novel usage of union and intersection of muses to achieve runtime efficiency. They proposed a novel maximal satisfiable subsets enumeration algorithm called RIME [24]. The experimental results showed that RIME is several times faster than existing tools. In [25], they focused on the enumeration of muses, and introduced a domain agnostic tool called MUST. This tool outperforms other existing domain agnostic tools and is even competitive to fully domain specific solutions.

Computing minimum unsatisfiable cores is the most difficult, time-consuming, the highest time and space complexity. Therefore there is little consideration on extracting the minimum unsatisfiable subformulas by the researchers. To the best of our knowledge, only a few existing works [4–6] can derive the minimum unsatisfiable cores. Because the branch-and-bound algorithm [4] and greedy-genetic algorithm [5] are the two optimal ones, they are selected to accelerate the predicate abstraction.

4   Predicate Abstraction of PicoJava Microprocessor

Since the emergence of the Java language, it has been received considerable attention due to its inherent advantages. The execution speed of Java bytecode is continued to increase because its execution mechanism is from interpreting to compiling. In addition, it also attempts to improve the execution performance of the Java language by designing a new type of microprocessor. Consequently, PicoJava is the microprocessor series designed by SUN to meet this requirement. The PicoJava microprocessors can support the Java virtual machine environment in hardware, and directly execute the Java virtual machine bytecode. PicoJava-II is the second generation microprocessor after PicoJava-I.

PicoJava-II is a 32-bit microprocessor, and supports 300 instructions. It includes a 6-stage integer pipeline, which are the fetching instruction stage, the decoding stage, the reading registers stage, the execution stage, the accessing memory stage and the writing back stage. The PicoJava microprocessor can flexibly configure the size of the data cache and instruction cache, and choose whether contains floating-point unit or not. The architecture of PicoJava-II is illustrated in Fig. 1. It mainly is composed of the following modules: the instruction Cache, the data Cache, the integer unit, the floating-point unit, the stack

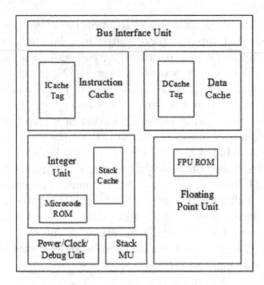

Fig. 1. Architecture of PicoJava-II

management unit, the power/clock/debug unit, and the bus interface unit, etc. Since the application paradigm of predicate abstraction is mainly based on the instruction Cache unit of PicoJava-II, the architecture and function of this module will be introduced as following.

The instruction Cache unit is primarily used to store the instructions in the software programs. The instruction Cache of PicoJava-II can be flexibly configured as 0 KB, 1 KB, 2 KB, 4 KB, 8 KB and 16 KB. The instruction Cache unit mainly consists of several modules such as the direct mapping memory banks, the 16-byte instruction buffer, the instruction Cache control logic, the instruction buffer control logic, and the instruction Cache data path, etc. This instruction Cache unit supports low power consumption mode. The architecture of instruction Cache unit is depicted in Fig. 2. The instruction Cache control logic fetches instructions from the Cache memory banks and sends them to the instruction fetch stage of the integer pipeline. The instruction buffer stores the instructions fetched from the memory banks until the integer pipeline reads them. If the instruction buffer is empty and the Cache is missed, the integer pipeline will stop. Then the instruction Cache unit will issue a memory reading request to the main memory outside.

While utilizing formal method to verify the assertions of the instruction Cache unit in the PicoJava-II microprocessor, it is necessary to abstract and refine the RTL-level codes of the circuit design. In the process of abstraction, a counterexample will be generated by the model checking tool. Then the simulation of the counterexample on the concrete model is converted into a formula φ = $ConcreteModel \wedge Counterexample$, so as to check whether the abstract counterexample is true or not. If the formula φ is unsatisfiable, it indicates that a false counterexample is produced. Then the unsatisfiable subformulas of φ need to be extracted. These unsatisfiable cores are used to eliminate false counterexamples during the refinement process. The smaller unsatisfiable subformulas are, the more false counterexamples are removed. Since the unsatisfiable cores can reduce the number of refined iterations, the minimum unsatisfiable cores extraction algorithms are adopted to accelerate the predicate abstraction of the circuits' verification. According to the construction rule of the formula φ, some formulas can be generated for the counterexample simulation process of each module in the instruction Cache unit, and are employed as the benchmark to evaluate runtime of two algorithms in the next section.

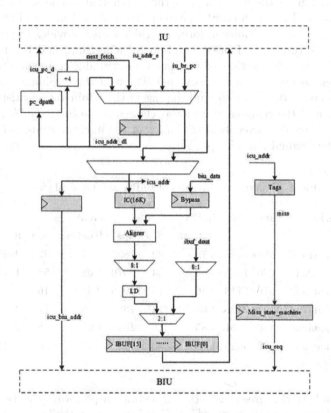

Fig. 2. Architecture of the instruction Cache unit

5 Evaluation Results and Analysis

To experimentally evaluate the efficiency between two algorithms of computing unsatisfiable cores on predicate abstraction of the circuits, we have selected a typical paradigm of the instruction Cache of the PicoJava-II microprocessor.

During verification of each module in the instruction Cache unit, some formulas were produced from the counterexamples simulation on the concrete model, taking advantage of the rules introduced in Sect. 4. These formulas are respectively named picojava_icu1.cnf through picojava_icu6.cnf. The branch-and-bound algorithm and the greedy-genetic algorithm are employed in the predicate abstraction process. The two algorithms are currently published fastest algorithms to derive the minimum unsatisfiable cores. The limit time was 1800 s. The experiments were conducted on a 2.5 GHz Athlon*2 machine having 2 GB memory and running the Linux operating system.

The experimental results of the branch-and-bound algorithm and greedy-genetic algorithm on the 6 formulas of the instruction Cache unit circuits are listed in Table 1. Table 1 shows the number of variables (vars) and the number of clauses (clas) for each Boolean formula. Table 1 also provides the runtime in seconds (BaBA runtime) of branch-and-bound algorithm, and the number of clauses (BaBA size) in the derived minimum unsatisfiable core, and the number of removed clauses per second (BaBA nps). The next three columns present the runtime in seconds (GGA time), and the size of the minimum unsatisfiable core (GGA size), and the number of removed clauses per second (GGA nps). The last column shows the percentage of clauses in the minimum unsatisfiable core occupying the original formula (per %).

Table 1. Experimental results of BaBA and GGA algorithm

Benchmarks	vars	clas	BaBA			GGA			Per
			Runtime	Size	nps	Runtime	Size	nps	(%)
picojava_icu1.cnf	561	1513	0.78	32	1899	0.61	32	2428	2.1
picojava_icu2.cnf	630	1645	1.64	54	970	1.08	54	1473	3.3
picojava_icu3.cnf	970	2718	1.36	16	1988	0.83	16	3255	0.6
picojava_icu4.cnf	1143	3128	1.65	18	1885	0.98	18	3173	0.6
picojava_icu5.cnf	1150	3151	1.85	18	1694	1.07	18	2928	0.6
picojava_icu6.cnf	1568	4520	17.5	30	257	14.3	30	314	0.7

From Table 1, the greedy-genetic algorithm outperforms the branch-and-bound algorithm, and has about 20%–40% reduction over the branch-and-bound algorithm on runtime. The greedy-genetic algorithm also has computed the exact minimum unsatisfiable subformula. The *nps* column represents the number of clauses removed by each algorithm in one second. It is computed by the equation $nps = (num-len)/time$, in which *num* denotes the number of clauses in the original formula, and *len* is the number of clauses in the minimum unsatisfiable cores, and *time* is the runtime of each algorithm in seconds. Then from the Table 1, the *nps* values of the greedy-genetic algorithm are much larger than the branch-and-bound algorithm. Therefore, the following conclusion can be reached that the runtime of greedy-genetic algorithm strongly exceeds the branch-and-bound algorithm, although the greedy-genetic algorithm occasionally obtains the nearly exact minimum unsatisfiable subformulas. Because the greedy-genetic algorithm achieves the best balance of computing efficiency and the unsatisfiable cores size, it remarkably outperforms the branch-and-bound algorithm on number of deleted clauses per second.

From the last column of Table 1, we may observe the following. For 6 formulas, the percentage of clauses in the minimum unsatisfiable cores is quite small, in most cases from 0.6% to 3.3%. In general, the minimum unsatisfiable cores can provide the most succinct explanations of infeasibility, and is very valuable for a variety of practical applications, such as the predicate abstraction. Therefore, in the process of predicate abstraction and refinement of the instruction Cache unit, the minimum unsatisfiable cores can help the tool to eliminate false counterexamples, and then minimize the number of refinement iterations.

From this typical application of the predicate abstraction and refinement, the advantages of unsatisfiable subformulas can be drawn as follows: by means of deriving the unsatisfiable cores, the causes of producing false counterexamples can be removed from the abstract model; furthermore the smaller unsatisfiable subformulas are, the more false counterexamples are pruned. Consequently, the greedy-genetic algorithm can compute the minimum unsatisfiable cores more efficiently, and then it can minimize the number of refinement iterations.

6   Conclusions

A minimum unsatisfiable core generally provides the most accurate explanation of infeasibility and is valuable for many applications. The predicate abstraction and refinement of microprocessor circuits is very difficult and time-consuming. Therefore, two algorithms of deriving minimum unsatisfiable cores, respectively called the greedy-genetic algorithm and branch-and-bound algorithm, are integrated into the predicate abstraction and refinement procedure. The instruction Cache unit of the PicoJava-II microprocessor is utilized as the benchmark. The results show that the greedy-genetic algorithm outperforms the branch-and-bound algorithm on runtime and removed clauses per second. We also analyzed that the unsatisfiable subformulas play a very important role in predicate abstraction and refinement of the circuits' verification.

References

1. Moskewicz, M.W., Madigan, C.F., Zhao Y., et al.: Chaff: engineering an efficient SAT solver. In: Proceedings of the 38th Design Automation Conference, pp. 530–535. ACM, Las Vegas, USA (2001)
2. Eén, N., Sörensson, N.: An extensible SAT-solver. In: Giunchiglia, E., Tacchella, A. (eds.) SAT 2003. LNCS, vol. 2919, pp. 502–518. Springer, Heidelberg (2004). https://doi.org/10.1007/978-3-540-24605-3_37
3. Jain, H., Kroening, D.: Word level predicate abstraction and refinement for verifying RTL Verilog. In: Proceedings of the 42nd Design Automation Conference, pp. 445–450. ACM, Anaheim, San Diego, USA (2005)

4. Liffiton, M.H., Mneimneh, M.N., Lynce, I., et al.: A branch and bound algorithm for extracting smallest minimal unsatisfiable formulas. Constraints **14**(4), 415–442 (2009)

5. Zhang, J.M., Li, S.K., Shen, S.Y.: Algorithms for Deriving minimum unsatisfiable Boolean subformulae. Acta Electronica Sinica **37**(3), 56–59 (2009)

6. Lynce, I., Marques-silva, J.: On computing minimum unsatisfiable cores, In: Proceedings of the 7th International Conference on Theory and Applications of Satisfiability Testing, pp. 305–310. Springer, Vancouver (2004)

7. Liffiton, M.H., Sakallah, K.A.: Algorithms for computing minimal unsatisfiable subsets of constraints. J. Autom. Reason. **40**, 1–30 (2008)

8. Gershman, R., Koifman, M., Strichman, O.: An approach for extracting a small unsatisfiable core. Formal Method Syst. Des. **33**(1), 1–27 (2008)

9. Gregoire, E., Mazuer, B., Piette, C.: Using local search to find MSSes and MUSes. Eur. J. Oper. Res. **199**(3), 640–646 (2009)

10. Nadel, A.: Boosting minimal unsatisfiable core extraction. In: Proceedings of 10th International Conference Formal Methods in Computer Aided Design, pp. 221–229. IEEE, Lugano, Switzerland (2010)

11. Ryvchin, V., Strichman, O.: Faster extraction of high-level minimal unsatisfiable cores. In: Sakallah, K.A., Simon, L. (eds.) SAT 2011. LNCS, vol. 6695, pp. 174–187. Springer, Heidelberg (2011). https://doi.org/10.1007/978-3-642-21581-0_15

12. Belov, A., Lynce, I., Marques-Silva, J.: Towards efficient MUS extraction. J. AI Commun. **25**(2), 97–116 (2012)

13. Nadel, A., Ryvchin, V., Strichman, O.: Efficient MUS extraction with resolution. In: Proceedings 13th International Conference Formal Methods in Computer Aided Design, pp. 197–200. IEEE, Portland, OR, USA (2013)

14. Belov, A., Heule, M.J.H., Marques-Silva, J.: MUS extraction using clausal proofs. In: Sinz, C., Egly, U. (eds.) SAT 2014. LNCS, vol. 8561, pp. 48–57. Springer, Cham (2014). https://doi.org/10.1007/978-3-319-09284-3_5

15. Bacchus, F., Katsirelos, G.: Using minimal correction sets to more efficiently compute minimal unsatisfiable sets. In: Kroening, D., Pǎsǎreanu, C.S. (eds.) CAV 2015. LNCS, vol. 9207, pp. 70–86. Springer, Cham (2015). https://doi.org/10.1007/978-3-319-21668-3_5

16. Bacchus, F., Katsirelos, G.: Finding a collection of MUSes incrementally. In: Quimper, C.-G. (ed.) CPAIOR 2016. LNCS, vol. 9676, pp. 35–44. Springer, Cham (2016). https://doi.org/10.1007/978-3-319-33954-2_3

17. Zhao, W., Liffiton, M.H.: Parallelizing partial MUS enumeration. In: Proceedings of IEEE 28th International Conference on Tools with Artificial Intelligence, pp. 464–471. IEEE, San Jose, CA, USA (2016)

18. Gregoire, E., Izza Y.: Boosting MCSes enumeration. In: Proceedings of the 27th International Joint Conference on Artificial Intelligence, pp. 1309–1315. ACM, Stockholm, Sweden (2018)

19. Narodytska, N., Bjorner, N., Marinescu, M.C., et al.: Core-guided minimal correction set and core enumeration. In: Proceedings of the 27th International Joint Conference on Artificial Intelligence, pp. 1353–1361. ACM, Stockholm, Sweden (2018)

20. Liu, S., Luo, J.: FMUS2: an efficient algorithm to compute minimal unsatisfiable subsets. In: Fleuriot, J., Wang, D., Calmet, J. (eds.) AISC 2018. LNCS (LNAI), vol. 11110, pp. 104–118. Springer, Cham (2018). https://doi.org/10.1007/978-3-319-99957-9_7

21. Luo, J., Liu, S.: Accelerating MUS enumeration by inconsistency graph partitioning. Sci. China Inf. Sci. **62**(11), 1–11 (2019). https://doi.org/10.1007/s11432-019-9881-0
22. Mencía, C., Kullmann, O., Ignatiev, A., Marques-Silva, J.: On computing the union of MUSes. In: Janota, M., Lynce, I. (eds.) SAT 2019. LNCS, vol. 11628, pp. 211–221. Springer, Cham (2019). https://doi.org/10.1007/978-3-030-24258-9_15
23. Bendík, J., Meel, K.S.: Approximate counting of minimal unsatisfiable subsets. In: Lahiri, S.K., Wang, C. (eds.) CAV 2020. LNCS, vol. 12224, pp. 439–462. Springer, Cham (2020). https://doi.org/10.1007/978-3-030-53288-8_21
24. Jaroslav, B., Cerna, I.: Rotation based MSS/MCS enumeration. In: Proceedings of the 23rd International Conference on Logic for Programming, Artificial Intelligence and Reasoning, pp. 120–137. EasyChair, Alicante, Spain (2020)
25. Bendík, J., Černá, I.: MUST: minimal unsatisfiable subsets enumeration tool. TACAS 2020. LNCS, vol. 12078, pp. 135–152. Springer, Cham (2020). https://doi.org/10.1007/978-3-030-45190-5_8

Algorithm Design Through the Optimization of Reuse-Based Generation

Haipeng Shi[1,3], Haihe Shi[2(✉)], and Shenghua Xu[1]

[1] School of Information Management, Jiangxi University of Finance and Economics, Nanchang 330032, China
[2] School of Computer and Information Engineering, Jiangxi Normal University, Nanchang 330022, China
haiheshi@jxnu.edu.cn
[3] School of Software, Jiangxi Normal University, Nanchang 330022, China

Abstract. It has been suggested that the notion of reuse be introduced to the formal process of generating algorithms from their specifications so that the programs can be developed automatically with low costs. Through studying algorithm generation under guidance of formal method we develop reuse-based techniques for reliable algorithm generation by reusing subsets of formal processes. We formalize the concepts and definitions for the approach via the language of category theory. The proposed method minimizes the process for the generation of a series of reliable algorithms for problems with linear structures, and transfers algorithm design tasks into routine work that can be implemented automatically. Furthermore, under the proposed framework, a series of algorithm variants for an ordered searching problem and a new sorting algorithm are generated, which demonstrate the effectiveness of the technique as well as its capacity to produce new advanced algorithms that have different structures from existing ones.

Keywords: Algorithm Generation · Formal method · Reuse-based optimization · Software development

1 Introduction

Computer software is required to be highly trustworthy due to pervasive, intensive, and critical computations. This need can only be met by developing more rigorous methodology for software development [1]. As the core of computer software [2], algorithm, especially its reliability and productivity, plays a critical role in both trustworthiness and application of software. Formal method and automation of algorithms have been shown to be important ways to improve the reliability and productivity of various algorithms. Many well-known computer scientists, such as T. Hoare, E. Dijkstra, R. Milner, A. Pnueli, D. Gries, etc. [3–7], have made significant con-tributions in this research area. As the Turing Award winner J. Gray points out in [8], automatic programming is the one of 12 important research goals in 21st century.

Software design relies on algorithm development that involves mathematical formulation, reasoning, and subtle techniques. Automatic programming must be built on a

© Springer Nature Singapore Pte Ltd. 2021
K. He et al. (Eds.): NCTCS 2020, CCIS 1352, pp. 14–32, 2021.
https://doi.org/10.1007/978-981-16-1877-2_2

solid framework in which algorithms can meet various requirements with reliability and desirable efficiency. However, the establishment of such a framework is by no means an easy task since algorithms can be highly mathematically complex. Formal method is particular kind of mathematically based techniques for the specification, development and verification of software systems, and has been shown to be an effective approach for automatic programming. For high-integrity systems involving safety and security, program generations require formal development and formal verification such as proofs of refinement from specifications. Thus the formal development/verification becomes vital in the automated generation of algorithm in applications. This is a challenging research area, and how to reduce the costs of formal development that can lead to the improvement of automatic generation of algorithms is still a main concern in software design in current literature.

Software reuse enables developers to leverage the past effort and facilitates significant improvements in software productivity and quality. It catalyzes improvements for productivity by avoiding redevelopment and allows us to incorporate the existing components whose reliability has already been established [9]. This idea can be extended to algorithm design with the support of mathematics. For instance, for most problems with linear structure, once a particular algorithm is formally specified and designed, parts of its design process may be reused by other algorithms so that the overall cost of algorithm development can be significantly reduced.

In this paper, we develop a formal framework for algorithm reuse that can optimize the process of algorithm generation and further improve the efficiency of the automatic generation. Based on the features of formal method in algorithm design, we establish the necessary underlying rules by formally developing a series of algorithm variants in terms of a class of important formal specifications. Its related concepts and definitions are formalized via the language of category theory. We further show that the proposed approach and technique can produce new algorithms different from existing ones and thus offer the flexibility in software design.

Different from current approaches in literature, our proposed approach provides a broader framework by establishing formal reuse-based program semantics from the viewpoint of optimization. A few related works are reviewed below.

Based on the Dijkstra-Hoare calculus [3, 4, 7], Franssen [10] developed an interactive programming tool, named Cocktail, that keeps track of all proof obligations, supports programmers for the proof construction and verifies the correctness of proofs and programs, and now it is mainly used in university education systems. Similar to algorithm calculus, Smith et al. [11] tried to implement automatic development from problem specification to executable code through specification refinement supported by category theory, and provided large reusable specification libraries and taxonomies of algorithm design theories for specification construction, specification refinement and program optimization. How-ever, there is no available effective guidance as well as procedure on how to select a proper algorithm design theory from libraries for a given problem. Moreover, how to incorporate the optimization rules for a specific program remains untouched.

By using generic technique and generalization, Yakhnis et al. [12] constructed a library of about 30 reusable generic algorithms in which, for a given problem, developers may only need to select a suitable generic algorithm from the library and adapt it through replacing its generic identifiers by conceptually similar data types and operations instead of directly using the program derivation techniques. Novak [13] allowed rapid specialization of generic procedures for applications through a computational mapping that de-scribes how a concrete type is implemented in an abstract form. Zheng et al. [14] presented four typical varieties of shuttle transportation problems where a supplier sends goods to a set of customers, returning to the source after each delivery, and designed a simple generic greedy algorithm that yields optimal solutions for all of them. Three generalization algorithms based on the von Neumann algorithm for linear programming are proposed in [15], and a C++ generic framework for modeling and solving convex optimization problems of variational image analysis is given in [16]. In terms of using generative programming approach, there has been some notable work, (see, e.g., [17] and references therein), which developed a system called Expression Code Generator that automatically generates the code for the mathematical formula. Nedunuri et al. combined problem specifications with flexible algorithm theories to derive efficient Haskell algorithms for three varieties of maximum segment sum problems [18]. However, a formal framework for the general case has not been seen yet.

In order to optimize system design by means of existing formal tools, Robidoux et al. [19] proposed a reliability markup language to formally describe dynamic reliability block diagrams (DRBD model) that defines a framework for modeling the dynamic reliability behavior of computer-based systems, and presented an algorithm that automatically converts a DRBD model into a colored Petri net, so that system properties of a DRBD model can be verified using an existing Petri net tool. In addition, Bolton et al. [20] developed the enhanced operator function model (EOFM) for describing task analytic models that capture the descriptive and normative human operator behavior, and presented an automated process for translating an instantiated EOFM into the model checking language, and then the system model can be verified using model checking in order to analyze properties related to the human-automation interaction, and they also summarize the works in this area [21].

Moreover, Liu [22] proposed a general and systematic design and optimization method, including three steps of Iterating, Incrementalizing and Implementing, for transforming clear specification of straight-forward computations into efficient implementations of sophisticated algorithms.

Our proposed optimization technique shares the similar motivation with the aforementioned works in the specification trans-formation, automated algorithm generation, or in the reuse methodology, but differ in the concerns and approach are quite different. We focus on the reuse of formal notions at a smaller granularity than those existing techniques and our method supports the auto-mated generation of most algorithms (recurrences) for problems with linear structure with a minimum cost. The approach is shown to be effective and is confirmed by the proposed prototype system. It demonstrates that the proposed approach substantially reduces the difficulty and the cost of formal algorithm development/generation. Moreover, the reliability of resultant algorithm is effectively guaranteed because of the formal and automatic support from the underlying software platform.

Sorting and searching are two fundamental problems in computer science. From a broader perspective, they make a benchmark case study of how to solve computer programming problems in general [23]. Yet little research on automated searching algorithms generation has been conducted so far. Different from the aforementioned works, we generate a variety of algorithms for a searching problem from a same formal specification, and produce a new algorithm for sorting problem. Although Clark *et al.* [24] have indicated early that by gaining experience about what choices are crucial for the derivation of an existing algorithm, new algorithms can be discovered by systematically investigating alternative derivation paths, current available results only can produce known algorithms. Under our framework, one can actually generate many new algorithms different from those existing ones, and thus the approach provides richer architectures in both algorithm development and software design.

The paper is organized as follows. In Sect. 2, we briefly introduce formal method and some basic notions of category theory that are relevant to our approach. Section 3 provides a formal framework of our proposed approach. The optimization technique is based on the partition selection for problems with linear structure in terms of algorithm reuse to lower the cost of algorithm generation. Under this framework we generate a series of algorithm variants for an ordered searching problem to illustrate the effectiveness of the proposed approach, and also produce a new efficient sorting algorithm to demonstrate its generative features. The paper ends with concluding remarks in Sect. 4.

2 Preliminaries

2.1 The Formal Method

A formal method specifies a given problem as formal specification rigorously by employing mathematical techniques, and makes the refinement process from a specification provable and derivable. Not only can it demonstrate the correctness and subtle ideas of algorithms, but also it plays an important role in revealing rules behind the algorithm design, and in exploring various automatic programming.

PAR (Partition-and-Recur) [25–29] is a practical formal approach for software development that effectively tackles the complexity of loop programs. It combines two powerful strategies that are often used in dealing with particular problems, i.e., partition and iteration, to form a unified and systematic approach, and covers a number of known algorithm design strategies, such as greedy algorithm, dynamic programming, di-vide-and-conquer method, enumeration, and so on. Hence it is an important methodology in automated algorithm generation.

As shown in Fig. 1, it consists of algorithm design language Radl, abstract programming language Apla, systematic algorithm design method, a set of rules for the specification transformation and a set of transformation tools called PAR platform.

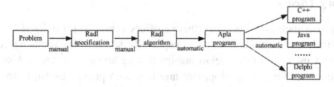

Fig. 1. Formal algorithm process based on PAR

Under the support of PAR platform, transforming Radl algorithms to Apla programs and transforming Apla programs to the programs of executable language are automatic.

The key concept to PAR is the recurrence relationship of the problem solving sequence, which is described in [25] as follows:

Definition 1 (Recurrence relation): Suppose that the solution of a problem P can be obtained by n computing steps which generate the solution sequence: $S_1, S_2,..., S_n$, where each S_i, $1 \leq i \leq n$, produced by one computing step, is a sub-solution of P, and S_n is the solution of P. Constructing a equation $S_i = F(\overline{S}_j)$, i.e. a sub-solution S_i is the function of sub-solutions $\overline{S}_j$, where $1 \leq i \leq n$, $1 \leq j < i$, and $\overline{S}_j$ stands for several sub-solutions. The equation $S_i = F(\overline{S}_j)$ is called recurrence relation of problem solving sequence, or recurrence for short.

In this setting, the process of developing an algorithm with its loop invariant for a given problem, as shown in Fig. 1, can be divided into five steps:

(1) Describe the formal functional specification of a problem using Radl.
(2) Partition the problem into a number of sub-problems, each of which has the same structure as the original problem but with a smaller size.
(3) According to the partition equation, formally construct the recurrence(s) from the formal functional specification.
(4) Generate the Radl algorithm as well as the loop invariant based on the recurrence and our new loop invariant strategies.
(5) Transform the Radl algorithm to Apla program based on the loop invariant, and further to the executable language program, say C++, Java, etc.

Most of specification transformation rules are quantifier properties. Let θ be a binary operator and Θ be the quantifier of operator θ, then $(\Theta i: r(i): f(i))$ means the quantity of $f(i)$ where i range over $r(i)$. The quantifier of binary operator $+$, $\wedge$, $\vee$, max(maximum),etc., are $\sum$, $\forall$, $\exists$, MAX, etc., respectively.

In general, provided that a problem can be divided into sub-problems and there is recurrence relationship among problem solving sequences, a reliable and efficient algorithm can be generated by PAR. This can be applied to develop many ingenious algorithms that have been shown to be convincing, such as Knuth's binary to decimal algorithm [26], complicated graph algorithms [27], cycle permutation [28], binary tree algorithms, combinatorial algorithms, etc., including the well-known algorithm of graph's planarity test and generating a planar one that was proposed by Turing Award winner J. Hopcroft and his student Tarjan.

2.2 Category Theory

As a "theory of functions", category theory offers a highly formalized language for software/algorithm specifications, and is especially suitable for focusing concern on reasoning about inter- and intra- relationships among software objects. Also, it is sufficiently abstract for formal development and has been proved useful in the semantic investigation for programming languages [30], which is proposed as a theory of data types and a framework for synthesizing formal specifications, e.g., see, Goguen [31]. Here we briefly introduce some categorical concepts that will be used later in the paper. More details can be found, for example, in [30].

Definition 2 (Category): A category C is.

- a collection Ob_C of objects
- a collection Mor_C of morphisms (arrows)
- two operations *dom, cod* assigning to each arrow f two objects respectively called domain and codomain of f
- an operation *id* (identity) assigning to each object b a morphism id_b such that $\text{dom}(id_b)$ $= \text{cod}(id_b) = b$
- an operation ∘ (composition) assigning to each pair f, g of arrows with $dom(f) =$ $cod(g)$ and arrow $f \circ g$ such that $dom(f \circ g) = dom(g)$, $cod(f \circ g) = cod(f)$ identity and composition must satisfy: (1) for any arrows f, g such that $cod(f) = b$ $= dom(g)$, we have $id_b \circ f = f$ and $g \circ id_b = g$; (2) for any arrows f, g, h such that $dom(f) = cod(g)$ and $dom(g) = cod(h)$, we have $(f \circ g) \circ h = f \circ (g \circ h)$

Definition 3 (Diagram): A diagram D in a category C is a directed graph whose vertices $i \in I$ are labeled by objects $d_i \in Ob_C$ and whose edges $e \in E$ labeled by morphisms f_e $\in Mor_C$.

Definition 4 (Cocone): Let D be a diagram in C, a cocone to D is.

- a C-object x
- a family of morphisms $\{f_i: di \to x \mid i \in I\}$ such that for each arrow $g: di \to dj$ in D, we have $fj \circ g = fi$, as shown in Fig. 2(a).

Definition 5 (Colimit): Let D be a diagram in C, a colimit to D is a cocone with C-object x such that for any other cocone with C-object x', there is a unique arrow $k: x \to x'$ in C such that for each d_i in D, $f_i: d_i \to x$ and $f_i': d_i \to x'$, we have $k \circ f_i = f_i'$, as shown in Fig. 2(b).

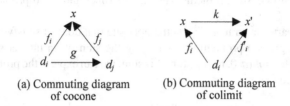

(a) Commuting diagram   (b) Commuting diagram
of cocone   of colimit

Fig. 2. Basic diagrams in category theory

Definition 6 (Functor): Let C and D be categories, a (covariant) functor $F: C \to D$ is a pair of operations $F_{ob}: Ob_C \to Ob_D$, $F_{mor}: Mor_C \to Mor_D$ such that, for each morphism $f: a \to b$, $g: c \to d$ in C,

- $F_{mor}(f)$: Fob$(a) \ \Box \ F_{ob}(b)$
- $F_{mor}(f \circ g) = F_{mor}(f) \circ F_{mor} (g)$
- $F_{mor}(id_a) = id_{Fob}(a)$

Definition 7 (Contravariant functor): Let C and D be categories, C^{op} be a category that reverses the directions of arrows in category C, a functor F: C^{op} D is called a contravariant functor.

3   Optimization Technique for Algorithm Generation

In this section, based on the characteristics of PAR in algorithm design, we present an optimization technique to generate algorithms from formal specifications by reusing the formal development process, and implement the proposed approach in a supporting platform.

3.1   Optimization Technique and Its Formal Model

According to PAR, a problem partition introduces a general algorithm design strategy by constructing a partition equation. It guides us to unfold and transform the problem specification by applying transformation rules to seek the relationship between solutions of original problem and its sub-problems. Obviously, different partitions correspond to different recurrences, and thus to different algorithms. Therefore it is the key design choice in algorithm development process, and is extremely crucial to the performance and style of the final algorithm.

For problems based on linear structure, there are often more than one feasible partition strategies. Generally, they can be categorized into two types: balanced and unbalanced, and the precise definition is given below.

Definition 8 (Balanced partition): Let $P(n)$ be a problem with size of n, balanced partition is to partition $P(n)$ into k sub-problems with equal size or differing by maximum of 1, $1 < k \leq n/2$. More specifically, supposing $n \bmod k = s$, $P(n)$ will be partitioned in a balanced way into k-s sub-problems with size of $\lfloor n/k \rfloor$ and s sub-problems with size of $\lceil n/k \rceil$.

Unbalanced partition strategy is just the opposite that doesn't satisfy Definition 8.

For example, given a problem of computing the sum of an integer sequence $a[0$: $n-1](n \geq 1)$, and let $sum(a, 0, n-1)$ store this sum, we can express the problem partition as the following equation:

$$sum(a, 0, n - 1) = F(sum(a, 0, \frac{n-1}{2}), sum(a, \frac{n-1}{2} + 1, n - 1)), n \geq 1$$

Here F is a function to be determined. Besides the above partition equation that implies a divide-and-conquer algorithm design strategy, we also have other alternative partitions that may lead to different design strategies, two of which are given below.

$sum(a, 0, n-1]) = F1(sum(a, 0, n-2), sum(a, n-1], n-1])), n \geq 1.$

sum(a, 0, n−1]) = F2(sum(a, 0, 0), sum(a, 1, n−1))), n ≥ 1.

According to the definition, sum(a, 0, n−2) is the sum of a[0: n−2]; if n == 1, a[0: −1] denotes an empty section and its sum is 0.

It may be critical to observe that a problem which can be partitioned in a balanced way can also be partitioned in an unbalanced way, i.e. unbalanced partition may be considered as a special form of balanced one. Because the ideas behind algorithms are captured naturally in terms of recurrences with PAR method, the intrinsic logic of algorithms may be similar as far as different partitions of a category are concerned. Hence, once the recurrence, i.e. algorithm corresponding to a balanced partition, has been derived from a problem specification, various algorithmic solutions with different performance for the problem can be generated by the reuse of that formal derivation process instead of carrying out a new formal derivation, and algorithm generation process thus can be optimized in terms of overall generation cost.

For example, suppose the F function has been determined, as listed below.

$$sum(a, 0, n - 1) = sum(a, 0, \frac{n - 1}{2}) + sum(a, \frac{n - 1}{2} + 1, n - 1), n \geq 1$$

The other two functions *F1* and *F2* can be worked out by constructing a mapping between the equations and the first one, respectively. It is easy to have the following two algorithms by different partitions.

sum(a, 0, n−1) = sum(a, 0, n−2) + sum(a, n−1, n−1]), n ≥ 1.

sum(a, 0, n−1) = sum(a, 0, 0) + sum(a, 1, n−1), n ≥ 1.

Taking advantage of rigorous formal derivation and automatic generation, the correctness of final algorithm (with respect to the specification) is guaranteed effectively.

By means of language of category theory, we further develop semantics for the technique, and make its description to be formal and precise. Thus the reliability of the proposed procedure can be guaranteed.

First we present the category theoretical semantics of problem specification and its solutions by defining problem category, problem solving category, and problem solving functor, respectively.

Definition 9 (Problem category): Given a problem structure P, P-SPEC is a category with problem specifications on P as objects, and a morphism between problem specification $p1$ and problem specification $p2$ is a mapping such that a morphism $f: p1 \rightarrow p2$ denotes $p2$ is a sub-problem of $p1$.

Obviously, morphisms in P-SPEC category satisfy composition. For any arrows f, $g \in Mor_{P\text{-}SPEC}, f: b \rightarrow c, g: a \rightarrow b$, we have $f \circ g: a \rightarrow c$, i.e., c is a sub-problem of a.

Definition 10 (Problem solving category): Given a problem structure P, P- RECUR is a category with the problem solving recurrences of P as objects, and a recurrence morphism between recurrence $r1$ and recurrence $r2$ is a calculation such that $f: r1 \rightarrow r2$, indicating that $r2$ can be obtained based on $r1$ through finite calculation steps.

For any arrows f, $g \in Mor_{P\text{-}RECUR}, f: r2 \rightarrow r3, g: r1 \rightarrow r2$, we have $f \circ g: r1 \rightarrow r3$, i.e., $r1$ is a sub-solution of $r3$. Therefore, the morphisms in P- RECUR category also satisfy composition.

Definition 11 (Problem solving contravariant functor): Given a problem structure *P*, *psf*: *P-SPEC* → *P-RECUR* is a problem solving contravariant functor by mapping problem specifications and their morphisms *p*1 → *p*2 of *P-SPEC* to the problem solving recurrences and recurrence morphisms *r*2 → *r*1 of *P-RECUR*.

Definition 12 (Problem partition functor): *npf*: *P-SPEC* → *P-SPEC* is a problem partition functor by mapping problem specifications and their partition morphisms *p*1 → *p*2 of *P-SPEC* to the problem specifications and partition morphisms *p*3 → *p*4 of *P-SPEC*.

The colimit operation can be used for constructing new recurrences mechanically. As shown in Fig. 3, it just need to compute colimit of *f* ∈ *psf* and *h* ∈ *npf* to optimize the process of algorithm generation from a problem specification by our technique, which is demonstrated as follows.

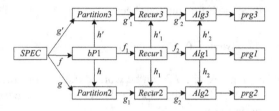

Fig. 3. Process optimization

Given a problem specification *SPEC*, we have derived one of its problem solving recurrences *Recur*1 by a balanced problem partition *bP*1, and generated its algorithm *Alg*1 based on *Recur*1. Suppose a new unbalanced specification partition *Partition*2 on *SPEC* gives the same number of sub-problems as *bP*1, instead of carrying out the new algorithm calculus of $g \circ g_1 \circ g_2$, *Recur*2 can be obtained by constructing a morphism *h*: *bP*1 → *Partition*2 and computing the colimit of *h* and the existing calculus f_1: *bP*1 → *Recur*1. And also the algorithm *Alg*2 can be obtained by computing the colimit of *h1*: *Recur*1 → *Recur*2 and the existing calculus f_2: *Recur*1 → *Alg*1. Similarly, for a new unbalanced partition *Partition*3, if there is a morphism *h*': *bP*1 → *Partition*3, then *Alg*3 can also be derived via the category theoretical computations. After the algorithms have been derived, they can be further refined into programs automatically, i.e., *prg*1, *prg*2, *prg*3, respectively.

Combined with the problem partition rules and specification transformation strategy that have been studied, we constructed an interactive Spec-Radl system that provides formal and semi-automatic support for the generation of Radl algorithms according to formal specifications. Its output is the input of our Radl-Apla program generation system, and then by the integrated program generation systems, such as Apla-Java, the Radl algorithm can be finally converted into executable programs.

And now we have extended our Spec-Radl system library with algorithm derivation of several classes of problems, such as searching, sorting, etc., and implemented this technique with C++ language through defining the following types and functions.

Part: <problem: String, num: Int>
Recur: String.
Map: <domain: String, range: String>
compPart: Part × Part → Map.
genRecur: Recur × Part × Part → Recur.

Here, *Part* type denotes problem partition; *Part.problem* stores sub-problems in the corresponding problem partition equation and *Part.num* stores the number of sub-problems; *Recur* type is used to denote recurrence, and *Map* type denotes mapping.

Additionally, function *compPart* is used to obtain the sub-problem mapping *h* between problem partitions *Partition*1 and *Partition*2, and function *genRecur* is used to generate the recurrence *Recur*2 according to a new partition *Partition*2.

Therefore, supported by the enhanced Spec-Radl system, an algorithmic solution may be derived step by step from a given formal specification formally and interactively, and algorithmic solutions for problems of the category may be generated automatically by our presented optimization technique. With further development, the algorithm derivation database can be extended and thus enable Spec-Radl system to generate more algorithms automatically in a way for minimizing the cost of algorithm generations.

3.2   Generation of Series of Ordered Searching Algorithms

One of the important prototypes of algorithms is searching algorithm that has been studied by many computer scientists (e.g., see [23] and references therein). It is motivated by a common belief that there are always some important features by studying how algorithms might be designed from the searching specifications. The searching problem is to look for a desirable element in a given set according to some specific key (perhaps containing other information related to the key).

Here we use our technique to formally develop series of algorithms for an ordered searching problem and demonstrate its effectiveness in our proposed approach.

Below two transformation rules used in the paper are presented.

(1)   Split with no overlap.
      $(\Theta\ i: r(i): f(i)) = (\Theta\ i: r(i) \wedge b(i): f(i))\ \theta\ (\Theta\ i: r(i) \wedge \neg b(i): f(i))$, where $b(i)$ is a boolean expression over i.
(2)   One point split.
      $(\Theta\ i: 0 \leq i < n: f(i)) = (\Theta\ i: 0 \leq i < n-1: f(i))\ \theta f(n-1)$.

The problem is to find an element *key* in a given sorted sequence $a[0:n-1]$. First we formally specify it as a decision problem. In fact, if necessary, we can obtain the location of *key* during the decision process.

$|[$sometype elem; **in** n: integer, *key*: elem, $a[0:n-1]$:list of elem; **out** b: boolean$]|$
AQ: $n \geq 1 \wedge (\forall i: 0 \leq i < n-1: a[i] \leq a[i+1])$;
AR: $b = (\exists k: 0 \leq k \leq n-1: a[k] == key)$;

Here *in* and *out* in front of precondition *AQ* are two key words defined in Radl and are used to denote the input and output respectively; *sometype* is another keyword denoting type parameter; *integer*, *boolean*, etc., are the predefined types in Radl and Apla, and AR stands for post-condition of algorithm.

Sequence is a predefined abstract data type that is implemented as *list* in Radl and Apla. Given a list variable S, $S[h]$, $S[t]$ and [] denote the head element of S, the tail element of S, and the empty list, respectively; $S[i]$ visits the ith element of S, and $S[i..t]$ denotes a sub-sequence of S.

The Binary Search Algorithm. Let $P(a, 0, n-1) = (\exists k: 0 \leq k \leq n-1: A[k] == key)$, it can be partitioned evenly as follows:

$p(a, 0, n-1]) \equiv F(p(a, 0, ((n-1])/2) -1), p(a, ((n-1])/2) + 1, n-1])), n \geq 1$.
Generally we have:

$$p(a, i, j) \equiv F(p(a, i, ((i + j)/2) - 1), p(a, ((i + j)/2) + 1, j)), 0 \leq i \leq j \leq n - 1. \tag{1}$$

Corresponding to the balanced partition, we can unfold problem specification to construct recurrence relation F as follows. Both explanations and transformation rules that are used are given in brace.

$p(a, i, j) = (\exists k: i \leq k \leq j: a[k] = key)$.
= {Split with no overlap}.
$(\exists k: i \leq k \leq ((i + j)/2)-1: a[k] = key) \vee (\exists k: k = (i + j)/2: a[k] = key) \vee (\exists k: ((i +j)/2) + 1 \leq k \leq j: a[k] = key)$.
= {One point split}.
$(\exists k: i \leq k \leq ((i + j)/2)-1: a[k] = key) \vee a[(i + j)/2] = key \vee (\exists k: ((i + j)/2) + 1 \leq k \leq j: a[k] = key)$.
= {Folding using definition of p}.
$p (a, i, ((i + j)/2)-1) \vee a[(i + j)/2] = key \vee p(a, ((i + j)/2) + 1, j)$.
<= {A known condition of AQ: $(\forall i: 0 \leq i < n-1: a[i] \leq a[i + 1])$}.
$(a[(i + j)/2] == key \wedge ((a[(i + j)/2] = key) \equiv true \vee (p(a, i, ((i + j)/2)-1) \vee p(a, ((i + j)/2) + 1, j)) \equiv false)) \vee (a[(i + j)/2] < key \wedge (p(a, i, ((i + j)/2)-1) \vee a[(i + j)/2] = key) \equiv false) \vee p(a, ((i + j)/2) + 1, j)) \vee (a[(i + j)/2] > key \wedge (p(a, i, ((i + j)/2)-1) \vee (a[(i + j)/2] = key \vee p(a, ((i + j)/2) + 1, j)) \equiv false))$.
= {Simplification}.
$(a[(i + j)/2] == key \wedge true) \vee (a[(i + j)/2] < key \wedge p(a, ((i + j)/2) + 1), j) \vee (a[(i +j)/2] > key \wedge p(a, i, ((i + j)/2) -1))$.
Therefore we have.

$$p(a, i, j) \equiv (a[(i + j)/2] == key \wedge true) \vee (a[(i + j)/2]$$
$$< key \wedge p(a, ((i + j)/2) + 1), j)) \vee (a[(i + j)/2]$$
$$> key \wedge p(a, i, ((i + j)/2) - 1)), 0 \leq i \leq j \leq n - 1. \tag{2}$$

It can also be rewritten in a mathematically transparent way as follows:

$$p(a, i, j) = \begin{cases} true, & a[(i+j)/2] == key \\ p(a, i, \dfrac{i+j}{2} - 1), & a[(i+j)/2] > key \\ p(a, \dfrac{i+j}{2} + 1, j), & a[(i+j)/2] < key \end{cases}$$

If $i == j == 0$, $a[i:((i+j)/2)-1)]$ and $a[((i+j)/2) + 1: j]$ denote empty sections, and $p(a, i, ((i+j)/2)-1)$ as well as $p(a, ((i+j)/2) + 1, j))$ are false.

Radl algorithm can be easily produced based on (2). And according to our new strategies for developing loop invariant [25], let a variable b store the searching result, we then can generate the following loop invariant mechanically based on recurrence.

ρ: $(b = true \land a[(i+j)/2] == key) \lor (b = p(a, i, ((i+j)/2)-1) \land a[(i+j)/2] < key) \lor (b = p(a, i, ((i+j)/2) -1) \land a[(i+j)/2] > key) \land 0 \leq i \leq j \leq n-1$.

The loop invariant clearly reveals the idea of a binary searching program. By using the Radl algorithm and loop invariant as input, code generator of PAR platform can generate the Apla program as well as the corresponding executable program automatically.

Next we will focus on generation of recurrences using our optimization technique, and spare the details of generation of final executable program which can be done automatically by our formulation.

The Ordered Linear Searching Algorithm. Alternatively, we can partition the original problem at its right boundary into two sub-problems with unequal size.

$$P(a, i, j) \equiv F(p(a, i, j - 1), p(a, j + 1, j)), \ 0 \leq i \leq j \leq n - 1. \tag{3}$$

Instead of carrying out new algorithm calculus, using our technique to construct a sub-problems mapping between (1) and (3), and the following recurrence corresponding to (3) can be achieved straightforwardly from (2).

$P(a, i, j) \equiv (a[j] == key \land true) \lor (a[j] < key \land p(a, j + 1, j)) \lor (a[j] > key \land p(a, i, j -1)), 0 \leq i \leq j \leq n-1$.

According to the definition of p predicate, we have $p(a, j + 1, j) \equiv false$, and the above recurrence can be simplified as.

$$p(a, i, j) \equiv (a[j] == key) \lor (a[j] > key \land p(a, i, j - 1)), \ 0 \leq i \leq j \leq n - 1. \tag{4}$$

It exactly embodies the idea of ordered linear searching algorithm that is similar to the one proposed by Knuth[23]. The important difference lies in that our algorithm is generated automatically through process reuse and optimization, governed by Radl algorithm and loop invariant.

The Other Three Linear Searching Algorithms. Given the following three unbalanced problem partition equations:

$$p(a, i, j) \equiv F(p(a, i, j - 2), p(a, j, j)), 0 \leq i \leq j \leq n - 1. \tag{5}$$

$$p(a, i, j) \equiv F(p(a, i, i-1), p(a, i+1, j)), 0 \le i \le j \le n-1. \tag{6}$$

$$p(a, i, j) \equiv F(p(a, i, i), p(a, i+2, j)), 0 \le i \le j \le n-1. \tag{7}$$

Note that $p(a, i, j-2) \equiv (\exists k: i \le k \le j-2: a[k] = key)$ with the condition $0 \le i \le j \le n-1$ according to the definition of p predicate, and it is false if i == j == 0 and a[i: j−2] is an empty section. The sub-problem $p(a, i+2, j)$ is similar.

It is easy to obtain the corresponding recurrences for the above three partition equations by applying our optimization technique via reusing formal derivation of binary searching algorithm. For partition Eq. (5),

$\dot{p}(a, i, j) \equiv (a[j-1] == key \wedge true) \vee (a[j-1] < key \wedge p(a, j, j)) \vee (a[j-1] > key \wedge p(a, i, j-2)), 0 \le i \le j \le n-1.$

And we have its recurrence (8) as follows by simplification.

$$p(a, i, j) \equiv (a[j-1] == key) \vee (a[j-1] < key \wedge a[j] == key)$$
$$\vee (a[j-1] > key \wedge p(a, i, j-2)), 0 \le i \le j \le n-1. \tag{8}$$

Similarly, we have the following recurrences for (6) and (7) respectively.

$$P(a, i, j) \equiv (a[i] == key \wedge true) \vee (a[i] < key \wedge p(a, i+1, j))$$
$$\vee (a[i] > key \wedge p(a, i, i-1)), 0 \le i \le j \le n-1.$$

$$\equiv (a[i] == key) \vee (a[i] < key \wedge p(a, i+1, j)), 0 \le i \le j \le n-1. \tag{9}$$

$$p(a, i, j) \equiv (a[i+1] == key \wedge true) \vee (a[i+1] < key \wedge p(a, i+2, j))$$
$$\vee (a[i+1] > key \wedge p(a, i, i)), 0 \le i \le j \le n-1.$$

$$\equiv (a[i+1] == key) \vee (a[i+1] < key \wedge p(a, i+2, j))$$
$$\vee (a[i+1] > key \wedge a[i] == key), 0 \le i \le j \le n-1. \tag{10}$$

Clearly, the asymptotic time complexities of these algorithms represented by recurrences (8), (9) and (10) approximate the ordered linear searching algorithm, that is, $O(n)$.

The Other Two Approximate Binary Searching Algorithms. Besides the balanced partition (1), the problem may also be partitioned into sub-problems as follows.

$$p(a, i, j) \equiv F(p(a, i, (i+j)/2), p(a, ((i+j)/2)+2, j)), 0 \le i \le j \le n-1. \tag{11}$$

$$p(a, i, j) \equiv F(p(a, i, i+((j-i+1)/3)-1), p(a, i+((j-i+1)/3)+1, j)), 0 \le i \le j \le n-1. \tag{12}$$

And their recurrences obtained by means of the optimization technique are listed below.

$$p(a, i, j) \equiv (a[((i+j)/2)+1] == key \wedge true) \vee (a[((i+j)/2)+1]$$

$$< key \wedge p(a, ((i + j)/2) + 2), j)) \vee (a[((i + j)/2) + 1]$$
$$> key \wedge p(a, i, (i + j)/2)), 0 \le i \le j \le n - 1. \tag{13}$$

$$p(a, i, j) \equiv (a[i + ((j - i + 1)/3)] == key \wedge \text{true}) \vee (a[i + ((j - i + 1)/3)]$$
$$< key \wedge p(a, i + ((j - i + 1)/3) + 1, j)) \vee (a[i + ((j - i + 1)/3)]$$
$$> key \wedge p(a, i, i + ((j - i + 1)/3) - 1)), 0 \le i \le j \le n - 1. \tag{14}$$

Above two algorithms are similar to the binary searching algorithm, but differ from the split position of the sequence. The performance of first algorithm corresponding to (13) is close to the binary searching algorithm, and the asymptotic time complexities of second one is $O(\log_{3/2} n)$.

3.3　Generation of a New Sorting Algorithm

The *sorting problem* presented in [23] can be formulated as follows. An array of n integer elements $a[0]$, $a[1]$,..., $a[n-1]]$ are given. We are to sort them into a non-decreasing order such that $a[i] \le a[i + 1]$ for $0 \le i < n-1$. Its formal specification may be described as follows.

|[**in** n: integer; **out** a[0:n−1]: array of integer; **aux** b[0:n−1]: array of integer]|
AQ1: $n \ge 0 \wedge a[0:n-1] == b[0:n-1]$;
AR1: $sort(a,0,n-1) \equiv ord(a,0,n-1) \wedge perm(a, b, 0, n-1)$;
$ord(a,0,n-1) \equiv (\forall k: 0 \le k < n-1: a[k] \le a[k + 1])$;
$perm(a,b,0,n-1) \equiv (\forall i: 0 \le i < n: (Nj: 0 \le j < n: a[j] = a[i]) = (Nk: 0 \le k < n: b[k] = a[i]))$;

Here, *aux* is a keyword in Radl, denoting the auxiliary variable(s); Auxiliary variable is used to establish the specification clearly and not allowed to be involved in algorithm, that is, as an auxiliary array, $b[0:n-1]$ cannot be computed in algorithm; N is a counting quantifier that satisfies:

$$(Ni : r(i) : f(i)) \equiv (\sum i : r(i) \wedge f(i) : 1)$$

Generally we have the definitions:

|[**in** n,i,j: integer; **out** a[i:j]:array of integer; **aux** b[i:j]: array of integer]|
AQ2: $n \ge 0 \wedge 0 \le i \le j < n \wedge a[i:j] == b[i:j]$;
AR2: $sort (a, i, j) \equiv ord(a, i, j) \wedge perm(a, b, i, j)$,
$ord(a, i, j) \equiv (\forall k: i \le k < j: a[k] \le a[k + 1])$,
$perm(a,b,i,j) \equiv (\forall k: i \le k \le j: (Ns: i \le s \le j: a[s] = a[k]) = (Nt: i \le t \le j: b[t] = a[k]))$;

(1) Problem partition.
The problem can be partitioned into three sub-problems. Generally we have.
$sort (a, i, j) \equiv F(sort(a, i + 1, j-1), sort(a, i, i), sort(a, j, j)) \wedge 0 \le i \le j < n$.
Further we can rewrite it as follows due to the single size of its last two sub-problems.
$sort (a, i, j) \equiv F(sort(a, i + 1, j-1), a[i], a[j]) \wedge 0 \le i \le j \le n-1$.
(2) Recurrence construction.
Below the recurrence relation F is constructed by means of specification transformation.

$ord(a, i, j) \equiv (\forall k: i \leq k < j: a[k] \leq a[k+1])$.
$\equiv$ {Split with no overlap and one point split}.
$a[i] \leq a[i+1] \wedge (\forall k: i+1 \leq k < j: a[k] \leq a[k+1])$.
$\equiv$ {Split with no overlap and one point split}.
$a[i] \leq a[i+1] \wedge a[j-1] \leq a[j] \wedge (\forall k: i+1 \leq k < j-1: a[k] \leq a[k+1])$.
$\equiv$ {Folding using definition of ord}.
$a[i] \leq a[i+1] \wedge a[j-1] \leq a[j] \wedge ord(a,i+1, j-1)$.
$=>$ {Let $Min(a, i, j)$ and $Max(a, i, j)$ denotes the minimum and maximum of $a[i:j]$ respectively, and $a[i] \leq a[i+1] \wedge ord(a,i+1,j-1) => a[i] = Min(a,i,j)$, $a[j-1] \leq a[j] \wedge ord(a,i+1,j-1) => a[j] = Max(a,i,j)$}.
$a[i] = Min(a,i,j) \wedge a[j] = Max(a,i,j) \wedge ord(a, i+1, j-1)$.

Through substituting this result into the definition of $sort$ (a,i,j), we have.

Recurrence 1.
$sort(a,i,j) \equiv a[i] = Min(a,i,j) \wedge a[j] = Max(a, i, j) \wedge sort(a, i+1, j-1), 0 \leq i \leq j \leq n-1$.

Here $sort(a, i+1, j-1)$ is a sub-problem satisfying specification {AQ3, AR3}, given briefly as follows:

I[**in** n,i,j: integer; **out** $a[i+1:j-1]$: array of integer; **aux** $b[i+1:j-1]$: array of integer]I
AQ3:$n \geq 0 \wedge 0 \leq i+1 \leq j-1 < n \wedge a[i+1:j-1] == b[i+1:j-1]$;

And $a[i] = Min(a, i, j)$ and $a[j] = Max(a, i, j)$ should be taken as two new problems to solve. Here we give the solutions directly.

Recurrence2.

$Max(a,i,j) \equiv \max(Max(a,i,(i+j)/2), Max(a, ((i+j)/2)+1, j)), 0 \leq i \leq j < n$.
$Min(a,i,j) \equiv \min(Min(a,i,(i+j)/2), Min(a, ((i+j)/2)+1,j)), 0 \leq i \leq j < n$.

(3) Loop invariant development.
Let two variables mi and ma store the minimum index and maximum index of $a[i:j]$ respectively, the following loop invariant can be obtained mechanically based on recurrences:
ρ: $sort(a,0,mi) \wedge sort(a,ma,n-1) \wedge a[ma] = Max(a,i,j) \wedge a[mi] = Min(a,i,j) \wedge 0 \leq i \leq j \leq n-1$.
Further we have the following loop invariant through unfolding ρ by substituting those predicate definitions into it.
$(\forall k: 0 \leq k < mi: a[k] \leq a[k+1]) \wedge (\forall k: ma \leq k < n-1: a[k] \leq a[k+1]) \wedge a[ma] = Max(a,i,j) \wedge a[mi] = Min(a,i,j) \wedge 0 \leq i \leq j \leq n-1$.
It shows the idea that sorts a by means of minimum selection and maximum selection simultaneously, and a consists of three sections at any time in the sorting process: two sorted sub-section located in two ends and one sorting sub-section in the middle.
(4) Program generation.
Apla program can be easily produced based on the recurrences and above loop invariant. Key code is given below.

i, j: = 0, n−1];
do i ≠ j → mi, ma: = i,j;
MaxMin(a, i + 1, j−1, mi, ma);
a[i], a[mi], a[j], a[ma]: = a[mi], a[i], a[ma], a[j];
i, j: = i + 1,j−1;
od;

Here *MaxMin(a, i + 1, j−1, mi, ma)* is a function, produced from the second recurrence, to search maximum and minimum of array $a[i + 1, j−1]$ simultaneously, and variable parameters *mi* and *ma* are used to store their indices respectively.

Further Java program can be generated from above Apla program by means of PAR platform.

(5) Algorithm complexity.

The time complexity of *MaxMin(0, n−1, max, min)* is $T(n) = (3n/2)−2$. The time complexity of the loop body of the program is $(3(n−2 * i)/2)−2$, hence the time complexity of above algorithm $T(n)$ is

$$\sum_{i=1}^{n/2-1} \left(\frac{3(n-2i)}{2} - 2\right) = \sum_{i=1}^{n/2-1} \left(\frac{3n}{2} - 2\right) - \sum_{i=1}^{n/2-1} 3 * i$$

$$= \left(\frac{n}{2} - 1\right)\left(\frac{3n}{2} - 2\right) - 3\left(\frac{n(n-2)}{8}\right) = \frac{3n^2 - 14n}{8} + 2$$

Obviously, its asymptotic time complexity is $O(n^2)$. However, it is slightly more efficient than those sorting algorithms whose time complexities are also $O(n^2)$ due to its constant factor, i.e., 3/8, is smaller than the constant factors of those algorithms, such as select sorting algorithm, insert sorting algorithm, bubble sorting algorithm, etc.

This is a new algorithm that comes from formal derivation and is unknown within the bound of our knowledge, and series of sorting algorithms can further be produced by reusing its formal derivation results.

3.4 Summary

Instead of constructing a new algorithm calculus, by applying our proposed technique, a series of ordered searching algorithms as well as sorting algorithms have been obtained with minimum costs. The generated algorithms include many variants that may contain different non-functionality properties. Each algorithm could be chosen to solve the corresponding problem under different scenarios. Clearly, it is also possible that the new algorithm that is not listed above may be generated by providing different partition equations. This approach can not only offer many new algorithms but also keep the total cost of algorithm generation as low as possible, and provides us the flexibility in software development at the same time. Figure 4 summarizes the generation of some searching algorithms, in which details of more morphisms are omitted.

The Radl algorithms can be easily established based on the above recurrences, and Apla programs can be generated from the algorithms. Moreover, executable programs can also be generated from Apla programs automatically, such as Java, C++/C#, Delphi,

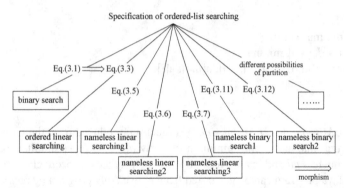

Fig. 4. Generation process for some of the searching algorithms by optimization (details of more morphisms are omitted)

etc., by making use of our PAR platform. We have transformed these Apla programs into Java programs by our Apla to Java system and their running results are desired, respectively.

Our approach, illustrated by the ordered searching problem and sorting problem, presents the idea of how the algorithm generation process can be optimized via reuse of algorithm (specification) refinements, and demonstrates the fact that the proposed method/technique is capable of producing new algorithms with little cost, rather than just verifying the correctness of previously known algorithms.

4 Conclusion

In this paper we present a reuse-based optimization technique for the reliable algorithm design of a class of linear-structural problems, and apply the category theoretical framework to formulate the desirable model. By applying the proposed approach, a series of ordered searching algorithms, including many new ones differing from those existing searching algorithms, as well as a new sorting algorithm have been generated automatically, and the mathematical setting of our framework can guarantee the reliability of the output codes with respect to various input specifications. Furthermore, an interactive algorithm design prototype system is also enhanced to support the application of our technique.

Our method can be used, according to the different user requirement with respect to time and space, to produce the different algorithms of a same problem. In order to generate a corresponding concrete program according to actual requirements, one just needs to decide how many sub-problems will be divided, and to select a preferred problem partition for the original problem. This enables a high-level of reusability and dynamic adaptability, and therefore makes the significant reduction of mathematical formulation required by formal development process, and thus the optimized procedure. This leads to the improvement of productivity and correctness of algorithms.

A large number of applications in practice demonstrate the favorable benefits by using the proposed optimization technique in this paper for algorithm development.

We have applied it to solve many algorithmic problems quite successfully, including those benchmark problems such as the permutation problems class, the searching problems class, the optimization problems class, etc. Our approach is also suitable to the trustworthy software fields including military support and reliable information systems.

Acknowledgments. This work was funded by the National Natural Science Foundation of China under Grant No.62062039, and the Natural Science Foundation of Jiangxi Province under Grant No.20202BAB202024.

References

1. Hoare, T.: The verifying compiler: a grand challenge for computing research. J. ACM. **50**(1), 63–69 (2003)
2. Harel, D., Feldman, Y.: Algorithmics: The Spirit of Computing, 3rd edn. Addison-Wesley, Upper Saddle River (2004)
3. Hoare, C.A.R.: An axiomatic basis for computer programming. Commun. ACM **12**(10), 576–585 (1969)
4. Dijkstra, E.W.: A Discipline of Programmin. Prentice-Hall Inc, Upper Saddle River (1976)
5. Milner, R.: Fully abstract models of typed lambda-calculi. Theor. Comput. Sci. **4**(1), 1–22 (1977)
6. Pnueli, A.: The temporal logic of programs. In: Proceedings of the 18th IEEE. FOCS, pp. 46–57 (1977)
7. Gries, D.: The Science of Programming. Springer, New York (1981). https://doi.org/10.1007/978-1-4612-5983-1
8. Gray, J.: What next? a dozen information-technology research goals. J. ACM. **50**(1), 41–57 (2003)
9. Selby, R.W.: Enabling reuse-based software development of large-scale systems. IEEE T. Software Eng. **31**(6), 495–510 (2005)
10. Franssen, M.: Cocktail: a Tool for Deriving Correct Programs. Ph.D. diss., Eindhoven University of Technology (2000)
11. Smith, D.R.: Generating Programs plus Proofs by Refinement. In: Meyer, B., Woodcock, J. (eds.) Verified Software: Theories, Tools, Experiments, pp. 182–188. Springer-Verlag, New York (2008). https://doi.org/10.1007/978-3-540-69149-5_20
12. Yakhnis, V.R., Farrell, J.A., Shultz, S.S.: Deriving programs using generic algorithms. IBM Syst. J. **33**(1), 158–181 (1994)
13. Novak, G.: Software reuse by specialization of generic procedures through views. IEEE T. Software Eng. **23**(7), 401–417 (1997)
14. Zheng, Y., Xu, C., Xue, J.: A simple greedy algorithm for a class of shuttle transportation problems. Optim. Lett. **3**(4), 491–497 (2009)
15. Gonçalves, J., Storer, R., Gondzio, J.: A family of linear programming algorithms based on an algorithm by von Neumann. Optim. Method. Softw. **24**(3), 461–478 (2009)
16. Breitenreichera, D., Lellmanna, J., Schnörra, C.: COAL: a generic modelling and prototyping framework for convex optimization problems of variational image analysis. Optim. Method. Softw. **28**(5), 1081–1094 (2013)
17. Idate, S., Patil, S.H., Mali, D.J.: Automated code generation using generative programming approach for a mathematical expression. In: Proceedings of the International Conference Systemics, Cybernetics and Informatics, pp. 134–137 (2008)

18. Nedunuri, S., Cook, W.R.: Synthesis of fast programs for maximum segment sum problems. In: Proceedings of the 8th International Conference on Generative Programming and Component Engineering, pp. 117–126 (2009)

19. Robidoux, R., Xu, H., Xing, L., Zhou, M.: Automated modeling of dynamic reliability block diagrams using colored petri nets. IEEE Trans. Syst. Man Cybern. A Syst. Hum. **40**(2), 337–351 (2010)

20. Bolton, M.L., Siminiceanu, R.I., Bass, E.J.: A systematic approach to model checking human-automation interaction using task-analytic models. IEEE Trans. Syst. Man Cybern. A Syst. Hum. **41**(5), 961–976 (2011)

21. Bolton, M.L., Bass, E.J., Siminiceanu, R.I.: Using formal verification to evaluate human-automation interaction: a review. IEEE Trans. Syst. Man Cybern. Syst. **43**(3), 488–503 (2013)

22. Liu, Y.A.: Systematic Program Design— from Clarity to Efficiency. Cambridge University Press, Cambridge (2013)

23. Knuth, D.: The Art of Computer Programming, vol. 3: Sorting and Searching, third ed. Addison-Wesley, Massachusetts (1998)

24. Clark, K., Darlington, J.: Algorithm classification through synthesis. Comput. J. **23**(1), 61–65 (1980)

25. Xue, J.: A unified approach for developing efficient algorithmic programs. J. Comput. Sci. Tech-CH. **12**(4), 314–329 (1997)

26. Xue, J., Davis, R.: A derivation and proof of Knuth's binary to decimal program. Software-Conc. Tool **18**, 149–156 (1997)

27. Xue, J.: Formal derivation of graph algorithmic programs using partition-and-recur. J. Comput. Sci. Tech-CH. **13**(6), 553–561 (1998)

28. Xue, J., Yang, B., Zuo, Z.: A linear in-situ algorithm for the power of cyclic permutation. In: Proceeding of the s 2nd International Frontiers. Algorithmics, pp. 113–123 (2008)

29. Xue, J., Zheng, Y., Hu, Q., et al.: PAR: a practicable formal method and its supporting platform. In: Sun, J., Sun, M. (eds.), ICFEM 2018. LNCS, vol. 11232. Springer, Cham (2018). https://doi.org/10.1007/978-3-030-02450-5_5

30. Asperti, A., Longo, G.: Categories, Types and Structures: An Introduction to Category Theory for the Working Computer Scientist. MIT Press, Cambridge (1991)

31. Goguen, J.A.: A categorical manifesto. Math. Struct. Comput. Sci. **1**, 49–67 (1991)

An Improved Firefly Algorithm for Software Defect Prediction

Lianglin Cao[1,2](✉) , Kerong Ben[1] , Hu Peng[2] , Xian Zhang[1] ,
and Feipeng Wang[1,2]

[1] Naval University of Engineering, WuHan 430033, China
charlies_navy@aliyun.com
[2] Jiujiang University, JiuJiang 332005, China

Abstract. Software defect prediction (SDP) plays an important role to help software development, where various advanced intelligence algorithms but no firefly algorithm (FA) are used to improve the prediction accuracy within a project or across projects. Current FA faces the problem of many unnecessary movements which reduces its efficiency of searching for an optimal solution. Therefore, an improved multiple swarms with different strategies firefly algorithm is proposed, named MSFA. The key principle of MSFA is to divide the swarm into three groups, where each group plays a different role to balance exploration and exploitation. Experimental studies were tested on CEC 2013 and SDP. The test results on CEC 2013 prove that MSFA achieves a high balance between the exploration and the exploitation. The test results conducted on SDP show that MSFA has a higher prediction accuracy, but much less computation cost compared with other FA variants.

Keywords: Firefly algorithm · Multi-swarm · Optimization · Software defect prediction

1 Introduction

Software defect prediction (SDP) is used to estimate the most defect-prone components of software by the most software engineers. To build a software defect prediction model, many classic algorithms are used as classifiers for SDP, such as logistic regression (LR) [1], decision table (DT) [2], support vector machine (SVM) [3], Bayesian Network (BN) [4]. Besides, some intelligence algorithms are employed for SDP, including the classic particle swarm optimization algorithm (PSO) [5], differential evolution algorithm (DE) [6], artificial bee algorithm (ABC) [7], ant colony optimization algorithm (ACO) [8], firefly algorithm (FA) [9], brain storm optimization algorithm (BSO) [10]. Although intelligence algorithms are simple in concept, easy to implement, and powerful in solving nonlinear, multidimensional, and complex optimization problems [11], the performance of the SDP employing these algorithms is not ideal. Based on these

Supported by the National Natural Science Foundation of China (No. 61763019).

K. He et al. (Eds.): NCTCS 2020, CCIS 1352, pp. 33–46, 2021.
https://doi.org/10.1007/978-981-16-1877-2_3

observations, an enhanced firefly algorithm (FA) is presented as an optimizer of SDP in this paper.

Firefly algorithm (FA) is a swarm intelligence algorithm. Due to its global search and intensive local search capabilities, it is widely applied in real-world applications. Recently, FA has been used to optimize software effort estimation models in [12]. However, few studies have applied FA to software defect prediction.

In FA, there is an attraction among any two fireflies. It may be unnecessary to introduce more attractions, where the bright fireflies are trapped in the local search. Therefore, how to reduce the redundancy of the search process is a fascinating topic. On the other hand, how the better fireflies move is often overlooked.

From the above investigations, a multiple swarms framework is suggested. The population is divided into three groups, each of which plays a different role in the search process, thus achieving a balance between exploration and exploitation. Also, MSFA is used to improve the prediction accuracy for SDP. Experimental results prove that MSFA is promising.

The main contributions of this paper are summarized as follows:

1. To improve the efficiency of FA, a framework for multiple swarms with different strategies is suggested.
2. A little-noticed of new research on how the better fireflies move is carried out.
3. The proposed approach is used to improve the accuracy of SDP with low time cost.

The rest of this paper is structured as follows: Sect. 2 introduces related work. The details of the proposed algorithm are described in Sect. 3. Section 4 provides the results of comparison with other improved FA variants and the results of comparison with other state-of-the-art intelligence algorithms for SDP. Finally, conclusions are drawn in Sect. 5.

2   Preliminary

2.1   Original Firefly Algorithm

FA is inspired by the flash characteristics of the fireflies, which is proposed by Yang (2009) [9]. Based on swarm intelligence, the simple concept is easy to implement, including three rules as follows:

1) Any two fireflies would introduce an attraction that is regardless of its gender.
2) The attractiveness is proportional to its brightness, and decreases as their distance increases. The weak firefly would move toward the brighter one. The brightest firefly would move randomly.

3) The fitness indicates the value of the objective function and is determined by the brightness of fireflies. For any fireflies x_i and x_j, the weak brightness x_i moves toward the brighter x_j, that movement is calculated by:

$$X_{id} = x_{id} + \beta_0 e^{(-\gamma r_{ij}^2)}(x_{jd} - x_{id}) + \alpha(rand() - \frac{1}{2}) \tag{1}$$

$$r_{ij} = \sqrt{\Sigma_{d=1}^d (x_{jd} - x_{id})^2} \tag{2}$$

x_{id} and x_{jd} is the dth dimension of firefly x_i and x_j in t^{th} iteration. X_{id} is the dth dimension of firefly x_i in $(t+1)^{th}$ iteration, which β_0 is the attractiveness when $r = 0$. γ is the light absorption coefficient set to 1. β_0 is set to 1. r_{ij} is the distance between x_i and x_j. a as well as $rand()$ is a random value between [0,1].

Algorithm 1. Firefly Algorithm

Initialize population and generate N fireflies;
Compute the fitness value of each firefly;
FEs=N, Input MaxFEs;
while FEs $\leq$ MaxFEs **do**
  for i=1 to N **do**
    for j=1 to i **do**
      if Light(i) > Light(j) **then**
        Update the new position of x_i according to Eq.(1);
        Evaluate the new solution;
      end if
    end for
  end for
  Rank the fireflies and find best value;
  t=t+1;
end while
Output the optimal solution;

2.2 Related Work

Since FA was proposed, many improved FA variants have been suggested. Wang et al. (2016) proposed a novel attraction model that random selected fireflies attract any other fireflies from the whole population, called random attraction firefly algorithm (RaFA) [13]. Its experimental results proved that the efficiency of FA was improved by reducing the unnecessary attractions. However, FA with less attractions may be poor in the global search.

Fister et al. (2012) suggested a perfect strategy on parameter level, named MFA [14]. It dynamically adjusts between parameter α and iteration t, which avoids trapping in the local search and obtains a good ratio between the exploration and exploitation, but ignores the efficiency of searching for an optimal solution.

Zhou et al. (2014) [15] introduced a role assignment scheme(RA) to enhance the performance of DE. In the proposed RA scheme, according to both the fitness and positional of individuals, the population was divided into some groups, and each played a different role in balancing the exploration and exploitation. Chen et al. (2019) [16]employed RA to improve BSO and obtained a good performance. In this paper, the population was divided by fitness information and location information, but unlike RA, they follow different grouping rules.

Other FA variants also achieved good performance. In [17], learning from the worst fireflies or the best fireflies to replace the worst fireflies, a new firefly was constructed, and the diversity of the population is successfully preserved. The slow convergence rate may be a disadvantage of it. In [18], to obtain a fast convergence rate, the Gaussian bare-bones approach was presented. Meanwhile, in [19], it suggested that all fireflies movements applied with the Gaussian distribution, but it has a high computational cost. A new method has been shown in [20], fireflies were divided into two groups by define gender. Male fireflies and female fireflies attract each other following different strategies.

Some algorithms were presented in hybrid level. A hybrid FA and PSO algorithm has been proposed by Berkan Aydilek in [21], it could exploit the advantage of both. Another hybrid FA and Cross-Entropy Method (CME) algorithm has been presented in [22]. Although its ability of the global search has been enhanced, the cost of CEM was still expensive. Additionally, a comprehensive analysis of FA various was introduced in [23,24].

From the above, most of the research is to achieve a balance between the exploration and exploitation. In FA, if the attraction model is changed, the structure of the swarm and the number of attractions will be changed, as well as the ability of the exploration and exploitation. Therefore, a novel multiple swarm attraction models is proposed. Where the numbers of the attraction become smaller, the exploration efficiency of search space is enhanced. To achieve a high-balance between the exploration and exploitation, three strategies are designed for the sub swarms.

3   Proposed Algorithm

3.1   Basic Idea

FA employs a full attraction model, attractions between any fireflies maybe not promising and redundant, which may reduce the efficiency of searching the optimal solution. Another reason is the use of fixed randomization parameters, which leads to a low balance between exploitation and exploration. To enhance the performance of FA, a multiple swarms with different strategies firefly algorithm (MSFA) is proposed in this paper. The framework is shown in Fig. 1.

Inspired by Zhou et al. (2014) [15] and Chen et al. (2019) [16], the population is divided into three groups, each of which plays a different role in the search process. These groups play three roles, called explorer, builder, and exploiter, respectively. The exploiters are expected to be individuals with better fitness. The objective of the exploiters is to avoid trapping in local search and find

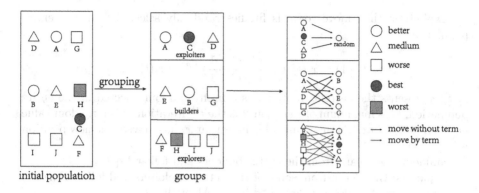

Fig. 1. The framework of MSFA.

better candidate solutions in their neighborhood. The explorers should be the individuals with worse fitness. The mission of the explorers is to explore new regions for some exploiters. The remaining population is the builders, whose goals are to explore new regions for themselves.

How to divide the population into three groups is as follows:

1. The average fitness of the whole population is calculated. It indicates as p_{mean}.

$$p_{mean} = \frac{1}{n} * \Sigma_{i=1}^{n} f_i \tag{3}$$

2. Two fireflies x_i and x_j are random selected firstly, Δf_i is computed by Eq. 4. It is used to represent the difference between the value fitness of the firefly and the average value of the whole population.

$$\Delta f_i = |f_i - p_{mean}| \tag{4}$$

If Δf_i is less than Δf_j, then firefly x_i is expected to be a builder. If the fitness of x_j is worse than p_{mean}, it is expected to be an explorer, otherwise, it is expected to be an exploiter.

3. Repeat 2 for the remaining individuals, until there are no fireflies to group.

In order to obtain a good ratio between the exploration and exploitation, each group employs a different strategy. Besides, the update equation of parameters α and β proposed in MFA [14] are adopted as follows:

$$x_i(t+1) = x_i(t) + \beta(x_j(t) - x_i(t)) + \alpha(t)\epsilon \tag{5}$$

$$\beta = \beta_{min} + (\beta_0 - \beta_{min})e^{(-\gamma r_{ij}^2)} \tag{6}$$

$$\alpha(t+1) = (\frac{1}{9000})^{\frac{1}{t}}\alpha(t) \tag{7}$$

The proposed strategies are presented as follows:

Exploiters: they move towards fireflies randomly selected from the entire population.

$$x_i(t + 1) = x_i(t) + \mu(\frac{G + t + 1}{t + G - 1})^{times_i}(x_i(t) - x_r) \tag{8}$$

$x_i(t)$ represents the current firefly, x_r is a firefly random selected in the entire population, G is maximum of iteration, t is the tth iteration, μ is a random value between $[0, 1]$, $times_i$ is the number of times that x_i has moved, the new position is $x_i(t + 1)$.

Builders: they learn from the better builders and follow Eq. (5).

Explorers: they learn from some of the better exploiters and follow Eq. (5).

The details of MSFA are described in the Algorithm 2.

Algorithm 2. Proposed MSFA

Input: sub swarm size k, number of movements of free firefly $times$, the maximum value of $times$ $Tmax$, conditions for iteration termination $MaxFEs$;
Initialize the population;
Compute the brightness of each firefly;
$FEs=N$;
while $FEs \leq MaxFEs$ **do**
    calculate p_{mean} and Δf_i;
    Divide the population into explorers, builders and exploiters;
    Compute the mean value of each group, $p_{explorers}, p_{builders}$ and $p_{exploiters}$
    if $p_{mean} \geq p_{builders}$ **then**
        Builders learning from itself, according to Eq. (8);
        Exploiters random walk, according to Eq. (5);
    else
        if $p_{explorers} \geq p_{exploiters}$ **then**
            Builders moves to the best individual, according to Eq. (5);
            Exploiters random walk, according to Eq. (8);
            Explorers moves to part of exploiters, according to Eq. (5));
        else
            Builders moves to part of exploiters, according to Eq. (5));
        end if
    end if
    Rank the fireflies and find best value;
    t=t+1;
end while
Output the best solution;

In the proposed algorithm, the exploiters with better fitness or better position, containing most of the better individuals, and maybe some of the medium individuals, which are expected to find the better candidate solutions in their neighborhood. The builders with medium fitness or position, including most of the medium individuals, maybe some of the better ones, and maybe some of the worse ones, which are expected to explore new regions. The explorers are made up of most the worse individuals, and perhaps some of the medium ones, which are expected to explore the search space of the exploiters. In that case, the diversity of a swarm is protected in each group, thus ensuring the sufficiency of exploring the search space. Also, not all groups working at the same time by setting conditions. Where the average fitness of the builders is better than that of the whole population, the exploiters and the builders get jobs. Otherwise, all the population take a mission. The attractions are reduced, thus the efficiency of FA is enhanced.

4   Experimental and Results

4.1   Experiment Preparation

To verify and analyze the performance of the proposed MSFA algorithm, a suit of well-known 28 benchmark functions called CEC2013 is used to test in the following experiment studies. More details about CEC2013 can be found in [28]. To ensure the fairness and objectiveness, all experimental studies are performed on a Laptop with Inter Core i7-9750 CPUS and 8 Gb RAM under Windows 10 platform. The proposed algorithm is compared with 5 FA versions. Besides, the parameters setting of the competitors involved is shown in Table 1.

Table 1. Parameters settings of compared algorithms

FA [9]	$\beta_0 = 1$, $\gamma = \frac{1}{\Gamma^2}$, $\alpha = 0.2$
DFA [26]	$\alpha = 0.5$
RaFA [13]	$\beta_{min} = 0.1$, $\beta_0 = 1$, $\alpha = 0.5$
NSRaFA [27]	$\beta_{min} = 0.3$, $\beta_{max} = 0.9$, $\gamma = \frac{1}{\Gamma^2}$, $\alpha = 0.5$
MSFA [14]	$\beta_{min} = 0.2$, $\beta_0 = 1$, $\gamma = 1$, $\alpha = 0.2$, $\mu = 0.2$

To get scientific results, each algorithm independently runs 30 times on CEC2013 problems with 30 dimensions. Also, two non-parametric statistical tests are used to analyze the numeric results, including the Wilcoxon test and Friedman test.

4.2   Comparison with FA Variants

Experimental results compared with several improved FA variants are shown in Table 2, including FA, DFA, RaFA, and NSRaFA algorithms. Besides, the number of individuals is set to 20, and the maximum of iterations is set to $5.0E + 5$.

To have statistically sound conclusions, the common tool Wilcoxon test is introduced to analyze the experimental results at a 0.05 significance level. The

Table 2. Experimental results of all algorithms on 30 dimensions

Function		FA	DFA	RaFA	NSRaFA	MSFA
f_1	MEAN	4.80E−02	2.79E−06	6.97E−13	3.51E+01	**6.52E−13**
	STD	4.54E−03	3.12E−07	2.75E−13	7.17E+00	**1.85E−13**
f_2	MEAN	**7.63E+04**	5.37E+06	1.70E+07	1.53E+07	1.09E+06
	STD	**3.27E+04**	2.26E+06	2.82E+06	3.35E+06	6.88E+05
f_3	MEAN	2.25E+07	1.08E+08	2.31E+09	4.07E+09	**1.35E+07**
	STD	2.44E+07	1.20E+08	1.52E+09	3.60E+09	**8.58E+06**
f_4	MEAN	**1.06E+03**	1.91E+04	1.55E+04	1.46E+03	2.40E+04
	STD	**6.41E+02**	5.13E+03	4.72E+03	5.34E+02	7.80E+03
f_5	MEAN	7.93E−02	1.14E+01	2.37E+01	1.12E+01	**1.15E−03**
	STD	1.04E−02	1.50E+01	3.73E+00	2.16E+00	**1.71E−04**
f_6	MEAN	**1.24E+01**	5.04E+01	8.25E+01	7.16E+01	2.79E+01
	STD	**1.69E+01**	2.77E+01	3.15E+01	4.59E+01	2.40E+01
f_7	MEAN	2.21E+07	1.93E+01	7.89E+01	1.18E+02	**2.63E+00**
	STD	7.21E+07	1.41E+01	1.78E+01	3.23E+01	**2.07E+00**
f_8	MEAN	**2.09E+01**	2.09E+01	2.10E+01	2.10E+01	2.10E+01
	STD	4.34E−02	**3.75E−02**	7.93E−02	4.31E−02	5.02E−02
f_9	MEAN	4.38E+01	1.28E+01	2.19E+01	2.70E+01	**9.87E+00**
	STD	2.96E+00	2.98E+00	4.03E+00	2.49E+00	**2.25E+00**
f_{10}	MEAN	6.28E−01	3.64E−02	3.04E+01	3.01E+01	**1.11E−02**
	STD	5.82E−02	3.22E−02	9.42E+00	6.79E+00	**9.48E−03**
f_{11}	MEAN	8.15E+02	3.81E+01	6.57E+01	1.24E+02	**3.30E+01**
	STD	2.06E+02	9.42E+00	1.67E+01	3.11E+01	**8.11E+00**
f_{12}	MEAN	7.05E+02	4.03E+01	6.45E+01	1.80E+02	**2.64E+01**
	STD	1.50E+02	1.70E+01	1.58E+01	4.65E+01	**1.56E+01**
f_{13}	MEAN	8.82E+02	8.12E+01	1.70E+02	1.94E+02	**5.65E+01**
	STD	1.24E+02	2.46E+01	3.65E+01	4.34E+01	**1.87E+01**
f_{14}	MEAN	4.78E+03	2.68E+03	3.74E+03	3.87E+03	**2.32E+03**
	STD	7.00E+02	7.55E+02	6.85E+02	**4.43E+02**	7.36E+02
f_{15}	MEAN	4.85E+03	2.40E+03	3.40E+03	4.72E+03	**2.20E+03**
	STD	6.14E+02	5.76E+02	1.15E+03	**5.43E+02**	7.24E+02
f_{16}	MEAN	4.00E−01	**2.32E−02**	2.61E−01	1.65E+00	3.98E−02
	STD	2.35E−01	**9.65E−03**	1.69E−01	3.34E−01	1.91E−02
f_{17}	MEAN	9.49E+02	6.83E+01	6.96E+01	2.14E+02	**6.34E+01**
	STD	1.46E+02	**9.68E+00**	1.03E+01	2.25E+01	1.33E+01
f_{18}	MEAN	8.94E+02	6.95E+01	7.31E+01	2.11E+02	**6.22E+01**
	STD	1.46E+02	1.75E+01	**1.19E+01**	1.84E+01	1.27E+01
f_{19}	MEAN	1.20E+02	**3.22E+00**	7.98E+00	1.53E+01	3.35E+00
	STD	1.34E+02	**7.50E−01**	4.62E+00	2.14E+00	7.70E−01
f_{20}	MEAN	1.50E+01	1.50E+01	1.49E+01	1.44E+01	**1.43E+01**
	STD	**8.89E−02**	0.00E+00	3.22E−01	7.75E−01	2.93E−01
f_{21}	MEAN	3.48E+02	3.48E+02	3.96E+02	3.64E+02	**3.18E+02**
	STD	8.07E+01	6.77E+01	6.63E+01	**4.61E+01**	8.20E+01
f_{22}	MEAN	6.83E+03	2.72E+03	4.54E+03	3.75E+03	**2.63E+03**
	STD	9.97E+02	8.09E+02	1.06E+03	**5.79E+02**	1.14E+03
f_{23}	MEAN	6.53E+03	3.22E+03	4.43E+03	5.47E+03	**2.59E+03**
	STD	7.56E+02	8.46E+02	1.05E+03	**7.47E+02**	1.11E+03
f_{24}	MEAN	5.22E+02	2.08E+02	2.43E+02	2.69E+02	**2.02E+02**
	STD	6.71E+01	**6.63E+00**	1.01E+01	7.53E+00	3.50E+01
f_{25}	MEAN	4.42E+02	**2.29E+02**	2.95E+02	2.65E+02	2.42E+02
	STD	2.12E+01	2.91E+01	1.33E+01	**1.00E+01**	3.15E+01
f_{26}	MEAN	3.62E+02	**2.00E+02**	2.65E+02	2.34E+02	2.67E+02
	STD	1.09E+02	**3.30E−02**	6.88E+01	6.73E+01	4.94E+01

(continued)

Table 2. (*continued*)

Function		FA	DFA	RaFA	NSRaFA	MSFA
f_{27}	MEAN	1.87E+03	**4.38E+02**	7.41E+02	1.01E+03	5.05E+02
	STD	1.44E+02	9.95E+01	9.95E+01	**8.00E+01**	9.86E+01
f_{28}	MEAN	5.29E+03	**3.00E+02**	9.01E+02	7.29E+02	3.08E+02
	STD	6.99E+02	**2.74E−03**	6.50E+02	6.42E+02	3.29E+02
Average Ranks		3.98	2.13	3.54	3.71	1.64
$+/\approx/-$			23/1/4	17/6/5	20/7/1	22/4/2

Wilcoxon's sum test results are indicated by "$+/\approx/-$", "+" means MSFA performs significantly better than the algorithm in the corresponding column, "$\approx$" means equivalent to, and "−" means worse than. The optimal solutions for each problem are shown in bold. One can observe that MSFA is significantly better than FA, DFA, RaFA, NSRaFA on 23, 17, 20, and 22 out of 28 benchmark functions, and worse than them on 4, 5, 1, and 2 problems, respectively. It indicates that MSFA not only improves the efficiency of the searching for an optimal solution, but also exploits a satisfied solution.

To analyze the stability of MSFA, the Friedman test is employed which is frequently applied in many fields. The smaller the value, the better the performance. From the results of the Friedman test, MSFA obtains average ranks of 1.64 as the minimal value from the comparison, and the result denotes that MSFA obtains the best performance in the comparison.

To give an intuitive comparison of these algorithms, Boxplots of the optimal solutions on 6 functions are shown in Fig. 2, including unimodal function(f_3, f_5), multimodal function(f_9, f_{13}) and composition function(f_{21}, f_{24}). The numeric

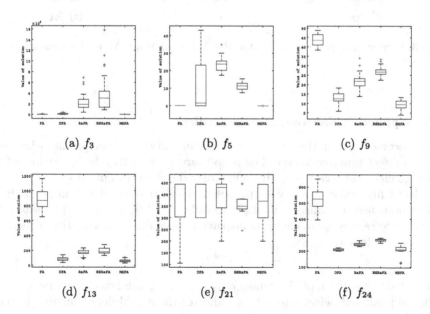

(a) f_3 (b) f_5 (c) f_9

(d) f_{13} (e) f_{21} (f) f_{24}

Fig. 2. Boxplots of FA, DFA, RaFA, NSRaFA, and MSFA for some selected functions

results come from Table 2. Compared with the competitors, MSFA achieves higher and more stable accuracy. It indicates that MSFA achieves high-balance between the exploration and exploitation.

To analyze the performance of MSFA furthermore, the convergence rate of all competitors are shown in Fig. 3. As we all known, if the MaxFEs value is the same, the better algorithm obtains both fast convergence rate and high accuracy value. From the results of convergence, MSFA is the better algorithm. It can be concluded that MSFA gets robustness and outperfors other competitors in the comparison.

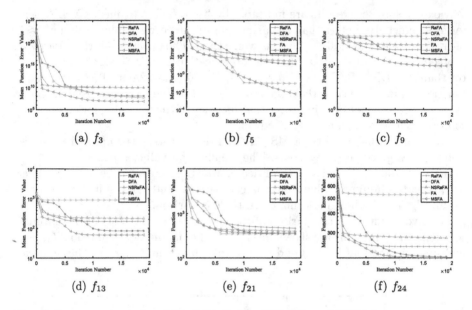

(a) f_3 (b) f_5 (c) f_9

(d) f_{13} (e) f_{21} (f) f_{24}

Fig. 3. Convergence graphs of FA, DFA, RaFA, NSRaFA, and MSFA for some selected functions

4.3   Applied in Software Defect Prediction

SDP becomes one of the important issues in software development, which is used to defect prone code according to software metrics, thereby improving software quality and safety [29]. In [30], CoDE used as an optimizer and a linear defect prediction model were proposed, which performed well in predicting defect-proneness rankings of multiple modules. In [31], a ranking performance measures were suggested, named fault-percentile-average. The equation of metric is defined as follows.

$$\frac{1}{k}\Sigma_{m=1}^{k}\frac{1}{n}\Sigma_{i=k-m+1}^{k}n_i \tag{9}$$

where k denotes the modules number, $f_1, f_2, \ldots, f_k$ are listed in an increasing order of predicted defect number, n_i indicates the actual defect number of the modules i, and $n = n_1 + n_2 + \ldots + n_k$ is the total number of defects. The larger fault-percentile-average, the better the performance.

Peng et al. (2019) [32] proposes neighbor-guided artificial bee colony algorithm(NABC) which applied in SDP, and obtains good performance. Based on these investigations, to verify the performance of MSFA over SDP, the metric fault-percentile-average is used. The larger value of testing means and training means indicates that the actual defects in software can be predicted more. Besides, the public benchmark datasets are employed, including five open-source software systems data - Eclipse JDT Core, Eclipse PDE UI, Equinox framework, Mylyn, and Apache Lucene, proposed by DAmbros et al. (2012) [33]. Compared with CoDE and NABC, the experiments are performed 10 times of tenfold cross-validation which conducted on the laptop with the MATLAB language, and the mean value of fault-percentile-average for 100 testing and training results are shown in Table 3.

Table 3. The experimental results of CoDE, NABC and MSFA for software detect prediction

Datasets	CoDE			NABC			MSFA		
	Training	Testing	Times(Sec)	Training	Testing	Times(Sec)	Training	Testing	Times(Sec)
JDT	0.8276	**0.8436**	33.7	0.8330	0.8363	33.9	**0.8333**	0.8377	33.8
PDE	0.7658	0.8650	70.2	0.7714	0.8695	71.4	**0.7724**	**0.8703**	70.1
Equinox	0.7933	0.6830	6.1	0.7958	**0.6970**	6.3	**0.7962**	0.6946	6.1
Lucene	0.8795	**0.7610**	18.5	**0.8860**	0.7530	19.3	0.8843	0.7557	18.7
Mylyn	0.7952	0.8168	101.4	0.7999	**0.8266**	103.5	**0.8002**	0.8215	103.1
Average ranks	2.6			1.9			1.5		

From the results of experiments, although the CPU time costs of all competitors are similar, the prediction accuracy for SDP is different. MSFA wins on training means of JDT, training means of PDE, testing means of PDE, training means of Equinox, and training means of Mylyn. CoDE wins on the testing mean of JDT and the testing mean of Lucene. NABC wins in the remaining datasets. While MSFA compared with CoDE, MSFA wins in 8 datasets except 2, it indicates that MSFA is significantly better than CoDE in SDP. While MSFA compared with NABC, MSFA wins in 7 out of 10 datasets, it denotes that MSFA is still better than NABC in the application of SDP. From the results of Friedman test, MSFA obtains average ranks of 1.5 as the best value from the comparison. Therefore, MSFA is the winner in the comparison.

To further verify the performance of MSFA for SDP, a comparison experiment of MSFA and two improved FA variants, RaFA and DFA, were carried out. The result of experiment is expressed in Table 4. MSFA not only obtains higher prediction accuracy, but also performs very well in terms of CPU times cost. The CPU time costs of MSFA is about one-third of RaFA, about one-thirtieth of DFA, respectively. It indicates that MSFA is much more effective than RaFA and DFA to solve SDP problem.

Table 4. The experimental results of RaFA, DFA and MSFA for software detect prediction

Datasets	RaFA			DFA			MSFA		
	Training	Testing	Times(Sec)	Training	Testing	Times(Sec)	Training	Testing	Times(Sec)
JDT	0.8336	**0.8418**	104.1	**0.8353**	0.8367	1464.2	0.8333	0.8377	**33.8**
PDE	0.7716	0.8688	206.7	0.7721	0.8701	2994.5	**0.7724**	**0.8703**	**70.1**
Equinox	0.7960	0.6926	18.1	0.7961	0.6903	193.1	**0.7962**	**0.6946**	**6.1**
Lucene	0.8856	0.7553	54.0	**0.8880**	0.7545	535.0	0.8843	**0.7557**	**18.7**
Mylyn	0.8005	0.8193	307.5	**0.8011**	0.8208	4193.8	0.8002	**0.8215**	**103.1**
Average ranks	2.2			2			1.8		

As far as we know, CoDE and NABC are the two state-of-the-art intelligence algorithms. The results of comparison with those indicate that MSFA has good performance in prediction accuracy of SDP. Compared with two improved FA variants, RaFA and DFA, MSFA is much more efficient than those in SDP. Thereby, MSFA is an effective and powerful new method for SDP.

5   Conclusion

Although FA is simple to implement and efficient in searching the optimal solution, it is easy to fall into a trap. In this paper, an improved multiple swarms with different strategies firefly algorithm is proposed, named MSFA. In the framework, the explorers, the exploiters, and the builders employ different strategies to play different roles in the search process. To verify the performance of MSFA, expensive experiments were conducted. All the experimental results proved that MSFA achieves a fit ratio between the exploration and exploitation. Furthermore, the application of MSFA in SDP also obtained good results. How to employ MSFA in solving other real-world application problems (Test case prioritization of software testing) would be our future work.

References

1. Bishop, C.-M., Nasrabadi, N.-M.: Pattern Recognition and Machine Learning. Springer, New York (2006)
2. Kohavi, R.: The power of decision tables. In: 8th European Conference on Machine Learning (ECML95), pp. 174–189. Heraklion, Crete, Greece (1995)
3. Lessmann, S., Baesens, B., Mues, C.: Benchmarking classification models for software defect prediction: a proposed framework and novel findings. IEEE. Trans. Softw. Eng. **34**(4), 485–496 (2008)
4. Okutan, A., Yıldız, O.T.: Software defect prediction using Bayesian networks. Empirical Softw. Eng. **19**(1), 154–181 (2012). https://doi.org/10.1007/s10664-012-9218-8
5. Eberhart, R., Kennedy, J.: A new optimizer using particle swarm theory. In: 1995 IEEE International Conference on Neural Networks, pp. 1942–1948, Nagoya, Japan (1995)

6. Peng, H., Guo, Z.-L., Deng, C.-S., Wu, Z.-J.: Enhancing differential evolution with random neighbors based strategy. Comput. Sci. **26**(1), 501–511 (2018)
7. Karaboga, D., Akay, B.: A comparative study of artificial bee colony algorithm. Appl. Math. Comput. **214**(1), 108–132 (2009)
8. Bell, J.E., McMullen, P.-R.: Ant colony optimization techniques for the vehicle routing problem. Adv. Eng. **18**(1), 41–48 (2004)
9. Yang, X.-S.: Firefly algorithms for multimodal optimization. In: Watanabe, O., Zeugmann, T. (eds.) SAGA 2009. LNCS, vol. 5792, pp. 169–178. Springer, Heidelberg (2009). https://doi.org/10.1007/978-3-642-04944-6_14
10. Peng, H., Deng, C.-S., Wu, Z.-J.: SPBSO: self-adaptive brain storm optimization algorithm with pbest guided step-size. Int. Fuzzy Syst. **36**(6), 5423–5434 (2019)
11. Fister, Jr., I., Fister, I., Yang, X.-S., Brest, J., : A comprehensive review of firefly algorithms. Swarm Evol. Comput. **13**, 34–46 (2013)
12. Ghatasheh, N., Faris, H., Aljarah, I., Al-Sayyed, R.-M.: Optimizing software effort estimation models using firefly algorithm. Softw. Engi. Appl. **8**, 133–142 (2018)
13. Wang, H., Wang, W., Sun, H., Rahnamayan, S.: Firefly algorithm with random attraction. Int. J. Bio Inspired Comput. **8**(1), 33–41 (2016)
14. Fister, Jr., I., Yang, X.-S., Fister, I., Brest, J.: Memetic firefly algorithm for combinatorial optimization. arXiv preprint arXiv:1204.5165 (2012)
15. Zhou, X., Wu, Z., Wang, H., Rahnamayan, S.: Enhancing differential evolution with role assignment scheme. Soft Comput. **18**(11), 2209–2225 (2013). https://doi.org/10.1007/s00500-013-1195-3
16. Chen, J.-Q., Deng, C.-S., Peng, H., Tan, Y., Zhou, X., Wang, F.: Enhanced brain storm optimization with role-playing strategy. In: 2019 IEEE Congress on Evolutionary Computation (CEC), pp. 1132–1139. Wellington, New Zealand (2019)
17. Yu, S., Zhu, S., Ma, Y., Mao, D.: Enhancing firefly algorithm using generalized opposition-based learning. Computing **97**(7), 741–754 (2015). https://doi.org/10.1007/s00607-015-0456-7
18. Peng, H., Peng, S.-X.: Gaussian bare-bones firefly algorithm. Int. J. Innova. Comput. Appl. **10**(1), 35–42 (2019)
19. Lv, L., Zhao, J.: The firefly algorithm with Gaussian disturbance and local search. J. Signal Process. Syst. **90**(8), 1123–1131 (2018)
20. Wang, C.-F., Song, W.-X.: A novel firefly algorithm based on gender difference and its convergence. Appl. Soft. Comput. **80**, 107–124 (2019)
21. Aydilek, I.B.: A hybrid firefly and particle swarm optimization algorithm for computationally expensive numerical problems. Appl. Soft. Comput. **66**, 232–249 (2018)
22. Li, G., Liu, P., Le, C., Zhou, B.: A novel hybrid meta-heuristic algorithm based on the cross-entropy method and firefly algorithm for global optimization. Entropy **21**(5), 494 (2019)
23. Tilahun, S.L., Ngnotchouye, J.M.T., Hamadneh, N.N.: Continuous versions of firefly algorithm: a review. Artif. Intell. Rev. **51**(3), 445–492 (2017). https://doi.org/10.1007/s10462-017-9568-0
24. Dey, N.: Applications of Firefly Algorithm and Its Variants. Springer, Singapore (2020)
25. Patwal, R.-S., Narang, N., Garg, H.: A novel TVAC-PSO based mutation strategies algorithm for generation scheduling of pumped storage hydrothermal system incorporating solar units. Energy **142**, 822–837 (2018)
26. Wang, J.: Firefly algorithm with dynamic attractiveness model and its application on wireless sensor networks. Int. J. Wire. Mob. Comput. **13**(3), 223–231 (2017)

27. Wang, H., Cui, Z., Sun, H., Rahnamayan, S., Yang, X.-S.: Randomly attracted fire-fly algorithm with neighborhood search and dynamic parameter adjustment mechanism. Soft Comput. **21**(18), 5325–5339 (2016). https://doi.org/10.1007/s00500-016-2116-z

28. Liang, J., Qu, B., Suganthan, P., Hernández-Díaz, A.-G.: Problem definitions and evaluation criteria for the CEC 2013 special session on real-parameter optimization. Comput. Int. Labo, Zhengzhou. Uni, Zhengzhou, CN. Nanyang. Techn. Uni, Singapore, Technical report. 201212(34), 281–295 (2013)

29. Peng, He., Li, B., Liu, X., Chen, J., Ma, Y.T.: An empirical study on software defect prediction with a simplified metric set. Int. J. Inf. Softw. Technol. **59**, 170–190 (2015)

30. Yang, X., Tang, K., Yao, X.: A learning-to-rank algorithm for constructing defect prediction models. In: Yin, H., Costa, J.A.F., Barreto, G. (eds.) IDEAL 2012. LNCS, vol. 7435, pp. 167–175. Springer, Heidelberg (2012). https://doi.org/10.1007/978-3-642-32639-4_21

31. Weyuker, E., Ostrand, T.-J., Bell, R.-M.: Comparing the effectiveness of several modeling methods for fault prediction. Int. J. Empiric Softw. Eng. **15**(3), 277–295 (2010)

32. Peng, H., Deng, C., Wu, Z.: Best neighbor-guided artificial bee colony algorithm for continuous optimization problems. Soft Comput. **23**(18), 8723–8740 (2018). https://doi.org/10.1007/s00500-018-3473-6

33. D'Ambros, M., Lanza, M., Robbes, R.: Evaluating defect prediction approaches: a benchmark and an extensive comparison. Int. J. Empiric Softw. Eng. **17**, 531–577 (2012)

Semi-online Algorithms for Hierarchical Scheduling on Three Parallel Machines with a Buffer Size of 1

Man Xiao[1], Lu Ding[1], Shu Zhao[1], and Weidong Li[1,2(✉)]

[1] School of Mathematics and Statistics, Yunnan University,
Kunming 650504, People's Republic of China
[2] Dianchi College of Yunnan University, Kunming 650000, People's Republic of China

Abstract. In this paper, we study the online problem on three hierarchical machines with a buffer size of 1, and have two hierarchy, the objective is to minimize the maximum machine load. When there is only one low-hierarchy machine, we give a low bound $\frac{3}{2}$ and present an online algorithm with competitive ratio at most $\frac{5}{3}$. When there are two low-hierarchy machines, we give a lower bound $\frac{3}{2}$ and present an online algorithm with competitive ratio at most $\frac{12}{7}$.

Keywords: Semi-online · Hierarchical machines · Competitive ratio · Buffer

1 Introduction

Given m hierarchical machines and n independent jobs, each job can be processed on some machines and each job only be processed by a machine. The hierarchical scheduling problem, with the objective is to maximize the minimum machine load [14,15].

When the objective is to minimize the makespan, denoted by $P|GoS|C_{max}$. Hwang et al. [1] investigated the offline problem $P|GoS|C_{max}$ and proposed a $\frac{5}{4}$-competitive algorithm for $m = 2$; also designed a $2 - \frac{1}{m-1}$-competitive algorithm for $m \geq 3$. Then, Ou et al. [2] designed a $\frac{4}{3}$-competitive algorithm and obtain a polynomial time approximation scheme (PTAS, for short) for $P|GoS|C_{max}$. For a special case of the problem $P|GoS|C_{max}$, Li et al. [3] presented an efficient polynomial time approximation scheme (EPTAS, for short) with running time $O(nlogn)$; for the problem $Pm|GoS|C_{max}$, and m is a constant, they also designed a fully polynomial time approximation scheme (FPTAS, for short) with running time $O(n)$.

Online scheduling over list is a scheduling that jobs arrive one by one, any new incoming job will be scheduled only according to the information of arrived jobs. An online algorithm's performance is measured by the competitive ratio. For any given instance I, let $C_A(I)$ (C_A, for short) be the objective value of algorithm A, for a minimization scheduling problem and a corresponding online algorithm A, the *competitive ratio* of the online algorithm A is defined as $\rho = \sup_I \frac{C_A(I)}{C_{OPT}(I)}$, i.e., $\rho C_{OPT}(I) \geq C_A(I)$ for any instance I, where $C_{OPT}(I)$ (C_{OPT}, for short) is the optimal offline objective value of the instance I. For the problem, if there is no online algorithm make the competitive

© Springer Nature Singapore Pte Ltd. 2021
K. He et al. (Eds.): NCTCS 2020, CCIS 1352, pp. 47–56, 2021.
https://doi.org/10.1007/978-981-16-1877-2_4

Table 1. Buffer size of 0 and 1

	One machine with hierarchy 1		Two machines with hierarchy 1	
	LB	UB	LB	UB
Buffer size of 0	1.801	1.857	1.824	1.857
Buffer size of 1	1.5	1.667	1.5	1.714

ratio less than ρ, the ρ is called a *lower bound* of this problem. If an online algorithm has a lower bound equal to its competitive ratio, it is called an optimal online algorithm.

For the online problem, Park et al. [4] and Jiang et al. [5] presented an optimal online algorithm for two hierarchical machines with competitive ratio of $\frac{5}{3}$ respectively. Lim et al. [6] presented an optimal online algorithm with a competitive ratio of 2 on three hierarchical machines. Wu et al. [7] designed some optimal semi-online scheduling algorithms for two hierarchical machines. Zhang et al. [8] first designed an online algorithm with a competitive ratio of $1 + \frac{m^2-m}{m^2-km+k^2} < \frac{7}{3}$ for any k and m, the k is the number of high-hierarchy machine, and got a competitive ratio of 1.857 for $m = 3$. For jobs have tightly-grouped processing time, Zhang et al. [9] designed some optimal online algorithms on the two hierarchical machines.

For the online problem $Pm|buffer|C_{max}$, Englert et al. [10] gave a lower bound of $\frac{3}{2}$ with a buffer size of $\lfloor \frac{m}{2} \rfloor - 1$, Lan et al. [11] gave a 1.5-competitive online algorithm with a buffer size of $1.5m$. When $m = 2$, Zhang [12] presented an optimal online algorithm with competitive ratio of $\frac{4}{3}$ and the buffer size of 1. For the online problem $P2|GoS, buffer|C_{max}$, Chen et al. [13] designed an optimal online algorithm with competitive ratio of $\frac{3}{2}$ and the buffer size of 1.

In this paper, we consider the online problem on three hierarchical machines with a buffer size of 1. There are two cases for this problem, when there is one machine of hierarchy 1 and two machines of hierarchy 2, we denote it as $P3(1,2,2)|buffer|C_{max}$. When there are two machines of hierarchy 1 and one machine of hierarchy 2, we denote it as $P3(1,1,2)|buffer|C_{max}$. Results of lower bound and the competitive ratio of buffer size of 0 and 1 can be found in Table 1. The rest of the paper is organized as follows. In Sect. 2, we define some basic notations. In Sect. 3, we consider the $P3(1,2,2)|buffer|C_{max}$, we give a lower bound $\frac{3}{2}$ and propose an online algorithm of competitive ratio at most $\frac{5}{3}$. In Sect. 4, we consider the $P3(1,1,2)|buffer|C_{max}$, we give a lower bound $\frac{3}{2}$ and propose an online algorithm of competitive ratio at most $\frac{12}{7}$. Finally, we make a summary.

2 Preliminaries

We are given three identical parallel machines M_1, M_2, M_3 and a job set $J = \{J_1, J_2, \ldots, J_n\}$. All jobs arrive online and all be scheduled irrevocably in their order of arrival. A new job arrives only after the current job is scheduled. We denote the j-th job in the arrival list as $J_j = (p_j, g_j)$, $p_j > 0$ is the *processing time*(also called the job's size) of the job J_j, and $g_j \in \{1, 2\}$ is the hierarchy of the job J_j. Each machine M_k has a hierarchy $g(M_k)$ and $g(M_k) \in \{1,2\}$, $k \in \{1,2,3\}$. For any job J_j, only can be processed by the machines which the hierarchy no more than it. p_j and g_j are not known until the job J_j arrives. If $g_j = i$, job J_j is called the job of hierarchy i, $i \in \{1,2\}$.

A schedule can be regard as a partition (S_1, S_2, S_3) of J, where S_k $(k = 1,2,3)$ contains the all jobs assigned to M_k. Let $L_k = \sum_{J_j \in S_k} p_j$ be the load of M_k, and T_i be the total processing time of the jobs of hierarchy i, $i = 1,2$. Our goal is to find a schedule such that $\max\{L_1, L_2, L_3\}$ is minimized.

For each $j \in \{1,2,\ldots,n\}$, $k \in \{1,2,3\}$ and $i \in \{1,2\}$, we define the following notions.

T_i^j: the total processing time of the first j jobs of hierarchy i.

L_k^j: the total processing time of the first j jobs scheduled on machine M_k after assigning job J_j.

p_{max}^j: the maximum processing time of the first j jobs of hierarchy 1.

q_{max}^j: the maximum processing time of the first j jobs of hierarchy 2.

B^j: the job in the buffer after scheduling the first j jobs.

For convince, we also let $L_k^n = L_k$, $T_i^n = T_i$, $p_{max}^n = p_{max}$ and $q_{max}^n = q_{max}$.

3 One Machine with Hierarchy 1

In this section, we consider the hierarchy of machine M_1 is 1, the hierarchy of machine M_2, M_3 is 2, and a buffer is available for storing at most one job. When a new job arrives, we have to assign it on a machine irrecoverably, or temporarily store it in the buffer. We give a lower bound $\frac{3}{2}$ and design an online algorithm with the competitive ratio at most $\frac{5}{3}$. In this section, for $j = 1,2,\ldots,n$, we let

$$LB^j = \max\{T_1^j, q_{max}^j, \frac{T_1^j + T_2^j}{3}\}. \tag{1}$$

Lemma 1. For the online problem $P3(1,2,2)|buffer|C_{max}$, the optimal makespan is at least LB^j after scheduled the first j jobs.

Proof. Let C_{OPT}^j be the optimal makespan after scheduled the first j jobs. For the online problem $P3(1,2,2)|buffer|C_{max}$, after the job J_j be scheduled, it is clearly that the $C_{OPT}^j \geq \max\{\frac{T_1^j + T_2^j}{3}, q_{max}^j\}$. Note that all jobs with hierarchy 1 only be processed on machine M_1, which implies $C_{OPT}^j \geq T_1^j$. So, the lemma holds.

Lemma 1 implies

$$C_{OPT} \geq \max\{T_1, q_{max}, \frac{T_1 + T_2}{3}\}. \tag{2}$$

Theorem 2. Any online algorithm A for $P3(1,2,2)|buffer|C_{max}$ has a competitive ratio at least $\frac{3}{2}$.

Proof. For any online algorithm A, the first two jobs in the sequence are $J_1 = (1,2)$ and $J_2 = (1,2)$, where the first number in brackets is the size and the second number is the hierarchy. We first consider algorithm A schedules the first two jobs to the machines. If A assigns both of them to the same machine, the last job $J_3 = (1,1)$ arrives, it's implies $C_A \geq 2$, $C_{OPT} = 1$. Else, the first two jobs are assigned to different machines, then, the last two jobs $J_3 = (2,1)$ and $J_4 = (2,2)$ arrive, then $C_A \geq 3$ and $C_{OPT} = 2$.

Next, we consider algorithm A stores a job in the buffer, without loss of generality, we consider A stores $J_1 = (1,2)$ into the buffer and assigns $J_2 = (1,2)$ to someone machine. If A assigns $J_2 = (1,2)$ to M_1, the last job is $J_3 = (1,1)$, then $C_A \geq 2$ and $C_{OPT} = 1$. If A not assigns $J_2 = (1,2)$ to M_1, the next job is $J_3 = (2,2)$, and we distinguish three cases for the storage of the buffer.

Case 1. $B^3 = \phi$, which means that A assigns first three jobs to the machines.
Case 1.1. A assigns $J_1 = (1,2)$ and $J_3 = (2,2)$ to the same machine.
The sequence stops, then $C_A \geq 3$ and $C_{OPT} = 2$.
Case 1.2. A does't assign $J_1 = (1,2)$ and $J_3 = (2,2)$ to the same machine.
If A assigns $J_3 = (2,2)$ to M_1, the last job is $J_4 = (2,1)$, then $C_A \geq 4$ and $C_{OPT} = 2$. If A assigns $J_3 = (2,2)$ to the machine which $J_2 = (1,2)$ is assigned in M_2 and M_3, the sequence stops, then $C_A \geq 3$ and $C_{OPT} = 2$. If A assigns $J_3 = (2,2)$ to the machine that is not assigned $J_2 = (1,2)$ in M_2 and M_3, and assigns $J_1 = (1,2)$ to M_1, the last job is $J_4 = (2,1)$, then $C_A \geq 3$ and $C_{OPT} = 2$. If A assigns $J_3 = (2,2)$ to the machine that is not assigned $J_2 = (1,2)$ in M_2 and M_3, and assigns $J_1 = (1,2)$ to the machine which $J_2 = (1,2)$ is assigned in M_2 and M_3, the last two jobs are $J_4 = (4,2)$ and $J_5 = (4,2)$, then $C_A \geq 6$ and $C_{OPT} = 4$, implies $\frac{C_A}{C_{OPT}} \geq \frac{3}{2}$.

Case 2. $B^3 = J_3 = (2,2)$, and A assigns $J_1 = (1,2)$ to someone machine.
Case 2.1. A assigns the $J_1 = (1,2)$ to M_1.
The last job is $J_4 = (2,1)$, then $C_A \geq 3$ and $C_{OPT} = 2$.
Case 2.2. A assigns $J_1 = (1,2)$ to the machine that is not assigned $J_2 = (1,2)$ in M_2 and M_3.
The last job is $J_4 = (2,2)$, then $C_A \geq 3$ and $C_{OPT} = 2$.
Case 2.3. A assigns $J_1 = (1,2)$ to the machine which $J_2 = (1,2)$ is assigned in M_2 and M_3.
Next job is $J_4 = (2,2)$, then one of $J_3 = (2,2)$ and $J_4 = (2,2)$ must be assigned, without loss of generality, we assume that $J_4 = (2,2)$ be assigned. If A assigns $J_4 = (2,2)$ to M_1, the last job is $J_5 = (3,1)$ and the sequence stops, then $C_A \geq 5$ and $C_{OPT} = 3$. If A assigns $J_4 = (2,2)$ to the machine that is not assigned $J_1 = (1,2)$ and $J_2 = (1,2)$ in M_2 and M_3, the last two jobs are $J_5 = (2,1)$ and $J_6 = (4,2)$, then $C_A \geq 6$ and $C_{OPT} = 4$. If A assigns $J_4 = (2,2)$ to the machine which $J_1 = (1,2)$ and $J_2 = (1,2)$ is assigned in M_2 and M_3, the sequence stops, then $C_A \geq 4$ and $C_{OPT} = 2$.

Case 3. $B^3 = J_1 = (1,2)$, and assigns $J_3 = (2,2)$ to someone machine.
Case 3.1. A assigns $J_3 = (2,2)$ to M_1.
The last job $J_4 = (2,1)$ arrives, then $C_A \geq 4$ and $C_{OPT} = 2$.
Case 3.2. A assigns $J_3 = (2,2)$ to the machine which $J_2 = (1,2)$ is assigned in M_2 and M_3.

The sequence stops, then $C_A \geq 3$ and $C_{OPT} = 2$.

Case 3.3. A assigns $J_3 = (2,2)$ to the machine that is not assigned $J_2 = (1,2)$ in M_2 and M_3. The next job $J_4 = (2,2)$ arrives.

Case 3.3.1. $B^4 = \phi$, which means that A schedules first four jobs to the machines. If A assigns $J_4 = (2,2)$ to M_1, and assigns $J_1 = (1,2)$ to the machine which $J_2 = (1,2)$ is assigned in M_2 and M_3. The last job is $J_5 = (3,1)$, then $C_A \geq 5$ and $C_{OPT} = 3$. Otherwise, the sequence stops, then $C_A \geq 3$ and $C_{OPT} = 2$.

Case 3.3.2. $B^4 = J_1 = (1,2)$, and A assigns $J_4 = (2,2)$ to someone machine. If A assigns $J_4 = (2,2)$ to M_1, the last job $J_5 = (3,1)$ arrives, then $C_A \geq 5$ and $C_{OPT} = 3$. If A not assigns $J_4 = (2,2)$ to M_1, the sequence stops, then $C_A \geq 3$ and $C_{OPT} = 2$.

Case 3.3.3. $B^4 = J_4 = (2,2)$, and A assigns $J_1 = (1,2)$ to someone machine. If A assigns $J_1 = (1,2)$ to M_1, the sequence stops, then $C_A \geq 3$ and $C_{OPT} = 2$. If A assigns $J_1 = (1,2)$ to the machine which $J_2 = (1,2)$ is assigned in M_2 and M_3, the last two jobs $J_5 = (2,1)$ and $J_6 = (4,2)$ arrive, then $C_A \geq 6$ and $C_{OPT} = 4$. If A assigns $J_1 = (1,2)$ to the machine which $J_3 = (2,2)$ is assigned in M_2 and M_3, the sequence stops, then $C_A \geq 3$ and $C_{OPT} = 2$.

Our algorithm is a modified version of the algorithm proposed in [13].

Algorithm H_1

Step 1. Set $j = 1$ and $L_1^{j-1} = L_2^{j-1} = L_3^{j-1} = 0$.

Step 2. If $g_j = 1$, assign J_j on M_1, and set $L_1^j = L_1^{j-1} + p_j$, $L_2^j = L_2^{j-1}$ and $L_3^j = L_3^{j-1}$.

Step 3. Else $g_j = 2$. Compare job J_j and the job that in the buffer(if no job in the buffer, regarding it as a job with size 0 in buffer). Put the bigger one job in buffer and denote it be B^j(B^j also indicate its size), and the smaller one be S^j (S^j also indicate its size).

    Step 3.1. If $L_2^{j-1} + S^j \leq \frac{5}{3}LB^j$, schedule S^j on M_2, and update L_i^j.

    Step 3.2. If $L_3^{j-1} + S^j \leq \frac{5}{3}LB^j$, assign S^j on M_3, and update L_i^j.

    Else, assign S^j on M_1, and update L_i^j.

Step 4. If there exist another job, set $j = j + 1$, return to **step 2.** Otherwise, assign B^j to the least loaded machine, and stop.

Theorem 3. ALGORITHM H_1 for $P3(1,2,2)|buffer|C_{max}$ has a competitive ratio at most $\frac{5}{3}$.

Proof. Proof by contradiction, we assume this theorem does not hold, then, we consider three cases according to which machine B^n is assigned to. Remember also using B^n to denote its size and $B^n = q_{max}$.

First, by algorithm H_1, we can get $L_2^{n-1} \leq \frac{5}{3}LB^n$ and $L_3^{n-1} \leq \frac{5}{3}LB^n$.

Case 1. B^n is assigned to M_1.

This means $L_1^{n-1} \leq \min\{L_2^{n-1}, L_3^{n-1}\}$ and $L_1 = L_1^{n-1} + q_{max}$, $L_2 = L_2^{n-1}$, $L_3 = L_3^{n-1}$. By the assumption of this theorem does not hold, we have $L_1^{n-1} + q_{max} > \frac{5}{3}LB^n$. Since $q_{max} \leq LB^n$, we have $L_1^{n-1} > \frac{2}{3}LB^n$. By the definition of LB^n, we have $T^n = L_1^{n-1} + L_2^{n-1} + L_3^{n-1} + q_{max} \leq 3LB^n$, so $L_2^{n-1} + L_3^{n-1} < \frac{4}{3}LB^n$, implies $\min\{L_2^{n-1}, L_3^{n-1}\} < \frac{2}{3}LB^n$. Then $L_1^{n-1} > \frac{2}{3}LB^n > \min\{L_2^{n-1}, L_3^{n-1}\}$, which contradicts with $L_1^{n-1} \leq \min\{L_2^{n-1}, L_3^{n-1}\}$.

Case 2. B^n is assigned to M_2. This means $L_2^{n-1} \leq \min\{L_1^{n-1}, L_3^{n-1}\}$ and $L_1 = L_1^{n-1}$, $L_2 = L_2^{n-1} + q_{max}$, $L_3 = L_3^{n-1}$.

Case 2.1. $L_2 \geq \max\{L_1, L_3\}$. If $L_2 = L_2^{n-1} + q_{max} > \frac{5}{3}LB^n$, since $q_{max} \leq LB^n$, we have $L_2^{n-1} > \frac{2}{3}LB^n$. Then by $T = T^n = L_1^{n-1} + L_2^{n-1} + L_3^{n-1} + q_{max} \leq 3LB^n = 3LB$, so $L_1^{n-1} + L_3^{n-1} < \frac{4}{3}LB^n$, implying that $\min\{L_1^{n-1}, L_3^{n-1}\} < \frac{2}{3}LB^n$, which contradicts with $L_2^{n-1} > \frac{2}{3}LB^n$ and $L_2^{n-1} \leq \min\{L_1^{n-1}, L_3^{n-1}\}$.

Case 2.2. $L_1 \geq \max\{L_2, L_3\}$.

If $L_1 > \frac{5}{3}LB^n$, since $T_1 = T_1^n \leq LB^n$, there must exist at least one job with $g_j = 2$ scheduled on M_1. Let job J_t be the last job of hierarchy 2 scheduled on M_1. Let T_1' be the total size of the jobs on M_1 after assigned job J_t, and the hierarchy of these jobs is 1. When job J_t is assigned on M_1, by algorithm H_1, we have $L_2^{t-1} + p_t > \frac{5}{3}LB^t$ and $L_3^{t-1} + p_t > \frac{5}{3}LB^t$, implies $L_2^{t-1} > \frac{2}{3}LB^t$ and $L_3^{t-1} > \frac{2}{3}LB^t$. Since $T^t = L_1^{t-1} + L_2^{t-1} + L_3^{t-1} + p_t + q_{max}' \leq 3LB^t$, so $L_1^{t-1} + q_{max}' < \frac{2}{3}LB^t$, implies $L_1^{t-1} + p_t < \frac{2}{3}LB^t \leq \frac{2}{3}LB^n$. Since $L_1 = L_1^n = L_1^{t-1} + p_t + T_1' > \frac{5}{3}LB^n$ and $T_1' \leq T_1 \leq LB^n$, we have $L_1^{t-1} + p_t > \frac{2}{3}LB^n$, which cause a contradiction with the $L_1^{t-1} + p_t < \frac{2}{3}LB^n$.

Case 3. B^n is assigned to M_3. This case is similar as the **Case 2**.

Thus, this theorem holds, we get $\frac{C_{H_1}}{C_{OPT}} \leq \frac{5}{3}$.

4 Two Machines with Hierarchy 1

In this section, we consider the hierarchy of machine M_1 and M_2 is 1, the hierarchy of machine M_3 is 2, and a buffer can temporarily store at most one job. When a new job arrives, we have to assign it on a machine irrecoverably; or temporarily store it in the buffer. We give a lower bound $\frac{3}{2}$ and design an online algorithm with the competitive ratio at most $\frac{12}{7}$. In this section, we let

$$LB^j = \max\{\frac{T_1^j}{2}, p_{max}^j, q_{max}^j, \frac{T_1^j + T_2^j}{3}\}, \tag{3}$$

clearly, $LB = LB^n$.

Lemma 4. For the online problem $P3(1,1,2)|buffer|C_{max}$, the optimal maximum machine load is at least LB^j after scheduled the job J_j for $j = 1, 2, \ldots, n$.

Proof. We denote the optimal maximum machine load after scheduled J_j as C_{OPT}^j. For the online problem $P3(1,1,2)|buffer|C_{max}$, after the job J_j be scheduled, it is clearly that the $C_{OPT}^j \geq \max\{\frac{T_1^j + T_2^j}{3}, p_{max}^j, q_{max}^j\}$. Note that all jobs with hierarchy 1 can be processed on machine M_1 and M_2, which implies $C_{OPT}^j \geq \frac{T_1^j}{2}$. So, the lemma holds.

Lemma 4 implies

$$C_{OPT} \geq LB = \max\{\frac{T_1}{2}, p_{max}, q_{max}, \frac{T_1 + T_2}{3}\}. \tag{4}$$

Theorem 5. Any online algorithm A for $P3(1,1,2)|buffer|C_{max}$ has a competitive ratio at least $\frac{3}{2}$.

Proof. For any algorithm A, the first three jobs in the sequence are $J_1 = (1,2)$, $J_2 = (1,1)$ and $J_3 = (1,1)$.

Case 1. $B^3 = \phi$, that means A assigns first three jobs to the machines.
If A assigns at least two jobs to M_1 or M_2, the sequence stops, we have $C_A \geq 2$, $C_{OPT} = 1$. Else, the last job $J_4 = (2,1)$ arrives, then $C_A = 3$, $C_{OPT} = 2$, implies $\frac{C_A}{C_{OPT}} \geq \frac{3}{2}$.

Case 2. $B^3 = J_1 = (1,2)$.
If A assigns $J_2 = (1,1)$ and $J_3 = (1,1)$ to M_1 or M_2, the sequence stops, then $C_A \geq 2$, $C_{OPT} = 1$. If A assigns $J_2 = (1,1)$ and $J_3 = (1,1)$ to M_1 and M_2 respectively, the last job $J_4 = (2,1)$ arrives, then $C_A \geq 3$, $C_{OPT} = 2$.

Case 3. $B^3 = J_2 = (1,1)$ ($B^3 = J_3 = (1,1)$ that is same as $B^3 = J_2 = (1,1)$).
Case 3.1. A assigns $J_3 = (1,1)$ and $J_1 = (1,2)$ to M_1 and M_2. The sequence stops, we have $C_A \geq 2$, $C_{OPT} = 1$.
Case 3.2. A assigns $J_3 = (1,1)$ to M_1 or M_2 and assigns $J_1 = (1,2)$ to M_3. Next job $J_4 = (2,2)$ arrives.
Case 3.2.1. $B^4 = \phi$.
If A assigns the $J_4 = (2,2)$ and at least one another job to the same machine, the sequence stops, we have $C_A \geq 3$, $C_{OPT} = 2$. Else, the last job is $J_5 = (1,1)$, we have $C_A = 3$, $C_{OPT} = 2$.
Case 3.2.2. $B^4 = J_2 = (1,1)$.
If A assigns $J_4 = (2,2)$ to machine that is assigned one job, sequence stops, we have $C_A \geq 3$, $C_{OPT} = 2$. If A assigns job $J_4 = (2,2)$ to the machine that is not assigned a job, the last job $J_5 = (1,1)$ arrives, then $C_A \geq 3$, $C_{OPT} = 2$.
Case 3.2.3. $B^4 = J_4 = (2,2)$.
If A assigns $J_2 = (1,1)$ to M_1 and M_2 that is not assigned job $J_3 = (1,1)$, the sequence stops, we have $C_A = 3$, $C_{OPT} = 2$. If A assigns the $J_2 = (1,1)$ to the machine which $J_3 = (1,1)$ is assigned in M_1 and M_2, the last job $J_5 = (1,1)$ arrives, then $C_A \geq 3$, $C_{OPT} = 2$.

The main ideal of algorithm H_2 is to schedule as many jobs with hierarchy 2 as possible to M_3, the details of algorithm H_2 are as follows.

Algorithm H_2

Step 1. Let $j = 1$ and $L_1^{j-1} = L_2^{j-1} = L_3^{j-1} = 0$.

Step 2. If $g_j = 1$, assign J_j to M_1 and M_2 which the load is the least, and update L_i^j.

Step 3. Else $g_j = 2$. Compare job J_j with the job in the buffer (if no job in buffer, regarding it as a job with size 0 in buffer). Put the bigger one job in buffer and denote it be B^j (B^j also indicate its size), and the smaller one be S^j (S^j also indicate its size).

Step 3.1. If $L_3^{j-1} + S^j \leq \frac{4}{7} T_2^j$, assign S^j on M_3, and update L_i^j.

Step 3.2. Else, assign S^j to M_1 and M_2 which the load is the least, and update L_i^j.

Step 4. When no new job arrives, if $L_3^{n-1} + B^n \leq \max\{L_1^{n-1}, L_2^{n-1}\}$, assign B^n on M_3; else, assign B^j to the least loaded machine, and stop.

Theorem 6. The competitive ratio of ALGORITHM H_2 for $P3(1,1,2)|buffer|C_{max}$ at most $\frac{12}{7}$.

Proof. Consider three cases according to which machine B^n is assigned to. We denote the total processing time of the jobs of hierarchy 2 that assigned to M_i as $L_i(2)$, for $i = 1, 2$, and remember $B^n = q_{max}$.

Case 1. B^n is assigned to M_3.

Case 1.1. $L_3 > \max\{L_1, L_2\}$, this means $L_3^{n-1} \leq \min\{L_1^{n-1}, L_2^{n-1}\}$ and $L_1 = L_1^{n-1}$, $L_2 = L_2^{n-1}$, $L_3 = L_3^{n-1} + q_{max}$. Since $L_3^{n-1} \leq \min\{L_1^{n-1}, L_2^{n-1}\}$, we have

$$C_{H_2} = L_3 = L_3^{n-1} + q_{max} \leq \frac{T_1 + T_2 - q_{max}}{3} + q_{max} = \frac{T_1 + T_2}{3} + \frac{2}{3} q_{max}.$$

Following (4), we have

$$C_{OPT} \geq \max\{\frac{T_1}{2}, p_{max}, q_{max}, \frac{T_1 + T_2}{3}\}.$$

So the competitive ratio $\frac{C_{H_2}}{C_{OPT}} \leq \frac{5}{3}$.

Case 1.2. $L_3 \leq \max\{L_1, L_2\}$.

If no jobs of hierarchy 2 be assigned to M_1 and M_2, then $L_3 = L_3^{n-1} + q_{max} = T_2$. By algorithm H_2, in this situation, the load between the M_1 and M_2 not exceed p_{max}, so

$$C_{H_2} = \max\{L_1, L_2\} \leq \frac{T_1 - p_{max}}{2} + p_{max} = \frac{T_1}{2} + \frac{p_{max}}{2}.$$

We get $\frac{C_{H_2}}{C_{OPT}} \leq \frac{3}{2}$.

Else, we let the last job of hierarchy 2 that be assigned to M_1 and M_2 is J_t, by algorithm H_2, we have $L_3^{t-1} + q_{max}^t \geq L_3^{t-1} + p_t > \frac{4}{7} T_2^t$. If there exist jobs of hierarchy 2 after J_t, them will all be assigned to M_3. Hence, $L_3 > \frac{4}{7} T_2$, implies $L_1(2) + L_2(2) < \frac{3}{7} T_2$.

By algorithm H_2, in any time, we know the load between the M_1 and M_2 not exceed $\max\{p_{max}, q_{max}\}$, so

$$C_{H_2} = \max\{L_1, L_2\} \leq \frac{3T_2/7 + T_1 - \max\{p_{max}, q_{max}\}}{2} + \max\{p_{max}, q_{max}\}$$

$$= \frac{3}{14}(T_1 + T_2) + \frac{2}{7}T_1 + \frac{\max\{p_{max}, q_{max}\}}{2}.$$

Following (4), we know $C_{OPT} \geq \max\{T_1/2, p_{max}, q_{max}, (T_1 + T_2)/3\}$, then, we get $\frac{C_{H_2}}{C_{OPT}} \leq \frac{12}{7}$.

Case 2. B^n is assigned to M_1. This means $L_1^{n-1} \leq \min\{L_2^{n-1}, L_3^{n-1}\}$ and $L_1 = L_1^{n-1} + q_{max}$, $L_2 = L_2^{n-1}$, $L_3 = L_3^{n-1}$.

Case 2.1. $L_1 \geq \max\{L_2, L_3\}$. Since $L_1^{n-1} \leq \min\{L_2^{n-1}, L_3^{n-1}\}$, by algorithm H_2, we get

$$C_{H_2} = L_1 \leq \frac{T_1 + T_2 - q_{max}}{3} + q_{max} = \frac{T_1 + T_2}{3} + \frac{2}{3}q_{max},$$

By (4), implying that $\frac{C_{H_2}}{C_{OPT}} \leq \frac{5}{3}$.

Case 2.2. $L_1 < \max\{L_2, L_3\}$. If $L_2 < L_3$, by algorithm H_2, we know $C_{H_2} = L_3 \leq \frac{4}{7}T_2$, following (4), we get $\frac{C_{H_2}}{C_{OPT}} \leq \frac{L_3}{T_2/3} \leq \frac{12}{7}$. If $L_2 \geq L_3$, we assume $L_2 > \frac{12}{7}LB$, because B^n is assigned to M_1 and the load between the M_1 and M_2 not exceed $\max\{p_{max}, q_{max}\}$, we have $L_1 \geq L_2 - \max\{p_{max}, q_{max}\}$. By algorithm H_2, because B^n is not assigned to M_3, we get $L_3 > L_2 - B^n = L_2 - q_{max}$. Hence,

$$T_1 + T_2 = L_1 + L_2 + L_3 > 3L_2 - \max\{p_{max}, q_{max}\} - q_{max}$$

$$\geq \frac{36}{7}LB - 2LB$$

$$= \frac{22}{7}LB$$

$$> T_1 + T_2.$$

Contradiction, so $L_2 \leq \frac{12}{7}LB$, and $\frac{C_{H_2}}{C_{OPT}} \leq \frac{L_2}{LB} \leq \frac{12}{7}$.

Case 3. B^n is assigned to M_2. This case is similar as the **Case 2**.

Hence, this theorem holds, obtain $\frac{C_{H_2}}{C_{OPT}} \leq \frac{12}{7}$.

5 Conclusion

We consider two cases for the online problem $P3|Gos, buffer|C_{max}$, but we can't get optimal online algorithms. It's valuable to design their optimal algorithms in future.

Acknowledgements. The work is supported in part by the Program for Excellent Young Talents of Yunnan University, Training Program of National Science Fund for Distinguished Young Scholars, IRTSTYN, and Key Joint Project of the Science and Technology Department of Yunnan Province and Yunnan University [No. 2018FY001(-014)].

References

1. Hwang, H., Chang, S.Y., Lee, K.: Parallel machine scheduling under a grade of service provision. Comput. Oper. Res. **31**(12), 2055–2061 (2004)
2. Ou, J., Leung, J.Y.T., Li, C.L.: Scheduling parallel machines with inclusive processing set restrictions. Nav. Res. Logist. **55**(4), 328–338 (2008)
3. Li, W., Li, J., Zhang, T.: Two approximation schemes for scheduling on parallel machines under a grade of service provision. Asia Pac. J. Oper. Res. **29**(05), 83–94 (2012)
4. Park, J., Chang, S.Y., Lee, K.: Online and semi-online scheduling of two machines under a grade of service provision. Oper. Res. Lett. **34**(6), 692–696 (2006)
5. Jiang, Y., He, Y., Tang, C.: Optimal online algorithms for scheduling on two identical machines under a grade of service. J. Zhejiang Univ. Sci. A **7**(3), 309–314 (2006)
6. Lim, K., Park, J., Chang, S.Y., Lee, K.: Online and semi-online scheduling of three machines under a GoS provision. In: Working Paper, Department of Industrial and Management Engineering, Pohang University of Science and Technology, Republic of Korea (2010)
7. Wu, Y., Ji, M., Yang, Q.: Optimal semi-online scheduling algorithm on two parallel identical machines under a grade of service provision. Int. J. Prod. Econ. **135**(1), 367–371 (2012)
8. Zhang, A., Jiang, Y., Tan, Z.: Online parallel machines scheduling with two hierarchies. Theoret. Comput. Sci. **410**(38), 3597–3605 (2009)
9. Zhang, A., Jiang, Y., Fan, L., Hu, J.: Optimal online algorithms on two hierarchical machines with tightly-grouped processing times. J. Comb. Optim. **29**(4), 781–795 (2015)
10. Englert, M., Ozmen, D., Westermann, M.: The power of reordering for online minimum makespan scheduling. In: 2008 49th Annual IEEE Symposium on Foundations of Computer Science, pp. 603–612. IEEE (2008)
11. Lan, Y., Chen, X., Ding, N., Han, X.: Online minimum makespan scheduling with a buffer. In: Proceedings of the Joint International Conference on Frontiers in Algorithmics and Algorithmic Aspects in Information and Management, pp. 161–171 (2012)
12. Zhang, G.: A simple semi on-line algorithm for $P2//C_{max}$ with a buffer. Inf. Process. Lett. **61**(3), 145–148 (1997)
13. Chen, X., Xu, Z., Dosa, G., Han, X., Jiang, H.: Semi-online hierarchical scheduling problems with buffer or rearrangements. Inf. Process. Lett. **113**(4), 127–131 (2013)
14. Li, J., Li, W., Li, J.: Polynomial approximation schemes for the max-min allocation problem under a grade of service provision. Discrete Math. Algorithms Appl. **1**(03), 355–368 (2009)
15. Xiao, M., Wu, G., Li, W.: Semi-online machine covering on two hierarchical machines with known total size of low-hierarchy jobs. In: Sun, X., He, K., Chen, X. (eds.) NCTCS 2019. CCIS, vol. 1069, pp. 95–108. Springer, Singapore (2019). https://doi.org/10.1007/978-981-15-0105-0_7

GDAssister: Graphic Design Assistant System with Optimal Algorithm of Associated Rule Mining

Dongyi Zheng[1], Haotian Wang[1], Panyu Liu[1], Fang Liu[2]([✉]), and Nong Xiao[1]

[1] College of Computer, National University of Defense Technology,
Changsha 410073, China
[2] School of Design, HuNan University, Changsha, China
`fangl@hnu.edu.cn`

Abstract. In the process of modern graphic design, designers usually get inspiration from a large number of art graphs and then choose graphic elements for graphic design. However, the process is generally manual, limiting the alternatives a designer can explore. In this paper, we propose GDAssister, a graphic design assistant system with optimal algorithm of associated rule mining. The system can automatically generate corresponding graphs according to the requirements of designers and use style transfer technology to change the style (color and texture) of the graphs. The system introduces four modules referring to the process of graphic design (e.g., graphic element extraction, association rule mining, automatic layout, style transform). Association rule mining is the core module of the system, and the efficiency of association rule mining algorithm determines the running efficiency of GDAssister. We design an improved association rule mining algorithm (IFP-Growth), which can reduce time consumption and memory usage. We use Chinese ethnic apparel database to verify the feasibility and efficiency of the system. Our experiment shows that GDAssister: (1)helps designers create more graphic alternatives to explore design creativity and (2)helps designers generate different styles of graphs more quickly to shorten the design cycle.

Keywords: Graph design · Graph elements extraction · Association data mining · Automatic layout · Style transform

1 Introduction

Graphic designers play a key role in contemporary social culture, and cultural industries related to graphic design account for an increasing proportion of the economy. According to the main responsibilities of the graphic designer, we divide the design process into four steps: the first step is to collect inspiration and materials, the second step is to select the basic graphic elements, and the third step is to form the first draft of the design by constantly adjusting the

Supported by organization nudt.

layout and style. Finally, the first draft is modified and fine-tuned to output the final draft. For most graphic designers, especially for novice graphic designers, the above process takes a lot of time and effort to explore alternatives. Although there are many kinds of design tools available, none of these tools can produce a large number of alternative designs in a short period of time.

With the gradual maturity of artificial intelligence image-related technology, more and more design tools begin to be infiltrated by artificial intelligence technology. The rapidly changing social needs often require the quickness of the design process. One of the main jobs for professional graphic designers is to organize several graphic elements to create different styles and themes. Our work is to design a graphics generation assistant system to support the possibilities of designing. We describe the main flow of current graphic design and analyze how designers make use of our proposed auxiliary system in the design process. In order to better assist designers in exploring graphic design, we propose GDAssister, to help designers explore the layout and style changes of image elements quickly. GDAssister can generate a variety of alternative layouts that can broaden the designer's vision. GDAssister includes the following four main modules:

- The graphic element extraction module uses content-based image retrieval technology, its purpose is to extract the basic elements contained in the Chinese ethnic apparel database, and then use these basic elements to build the basic element database, so as to provide data support for association rule mining.
- Mining association rules between basic elements and skeleton module uses the association rule mining algorithm, and then analyzes the relationship between skeletons and elements based on the graphic skeleton in the Chinese ethnic apparel database and the basic elements database. This module provides a creative source for designers and can recommend associated elements or skeletons according to the designer's input. The general mining process (e.g., Apriori, FP-Growth) consume large time and memory. Therefore, we design an improved association rule mining algorithm, which reduces the time consumption and memory usage in the mining process.
- According to the input of designer and the association between elements and skeletons, the automatic layout module produces a variety of graphics for designers to choose, and the designs of the layout module are used as the input of the style transform module.
- The style transform module generate style transform models that represent different styles to render the results of the layout, which can change the color and texture of the original image and provide inspiration for designers.

2   Related Work

The concept of association rule mining was first put forward in 1993, and it is widely used in various industries. The technology of association rule mining consists of two steps: finding all frequent itemsets and generating association rules.

It can find hidden value information from a large number of fuzzy, incomplete and random data. The representative algorithms are Apriori proposed by Agrawal et al. and FP-Growth proposed by J. Han et al. [1,11]. The whole mining process of FP-Growth only needs to scan the original data set twice, which improves the efficiency compared with Apriori. However, in practical application, these traditional association rule mining algorithms will produce redundant frequent item sets of worthless association rules.

The content-based image retrieval technology gives the expression and similarity measurement of image content for automatic processing, which overcomes the defects faced by using text in image retrieval, and gives full play to the advantage that the computer is good at computing [6,16]. In the method of feature extraction, content-based image retrieval can extract traditional features, such as SIFT, SUP, HOG, and other invariant local features with good anti-interference [19], and can also extract features of a certain layer of CNN classification model, such as VGG, ResNet, Inception and other classification models.

The traditional nonparametric image style transfer methods are mainly based on physical model rendering, texture, and synthesis. Efros et al. proposed a simple texture algorithm to synthesize a new texture by stitching and reassembling the sample texture [8]. With the rise of deep learning, Gatys et al. creatively proposed a kind of image style transfer based on convolution neural network [3,21]. At present, the application of image style transfer based on deep learning in intelligent graph design mainly has the following two main directions.

(1) Image processing. The traditional image processing technology intelligently processes the image with a fixed graph, while the emergence of image style transfer based on deep learning brings more imagination space for image style design and puts forward a content-aware style transfer method. Based on this method, we can use it to repair incomplete or blurred art images [9]. Zhang and others proposed a method of coloring cartoon sketches, which can be used to assist designers [5].

(2) An auxiliary tool for style design. Image style transfer can be used as a useful auxiliary tool, such as painting creation, architectural design, clothing design, game scene design and so on. Shengchao Li of Nanjing University has studied the application of deep learning in the rendering of image ink style and achieved good results [17]. Through the style transfer algorithm, Sun Dong et al. put forward a method that transfer the painting elements of Van Gogh's classic works to the film to form a unique animation artistic effect and verify the feasibility and scientific nature of the artificial intelligence style transfer algorithm applied to animation special effects design [18]. Jialin Ai has implemented a Web application based on B/S mode, in which designers can upload their own pictures. After selecting a style picture, the program will render the picture in the background and return the results to the designers [13].

3   The Design of GDAssister

GDAssister includes two databases: graphic elements and basic element database, and has four main functional modules (graphic element extraction, association rule mining, automatic layout, style transfer). Figure 1 shows the modules' interrelationship of GDAssister.

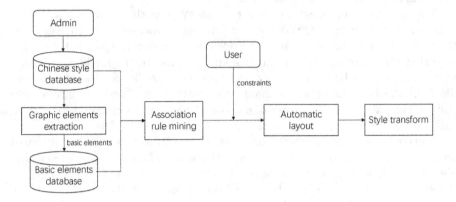

Fig. 1. Moudels of GDAssister.

(1) Chinese ethnic apparel database.

The Chinese ethnic apparel database is responsible for managing the cultural information, which is not only the knowledge source of association rule mining but also the material source of the graphic element extraction module. It mainly includes basic information such as type, function, format, size, year, author, collection place, skeleton, color matching. The database provides the administrator with the operation of adding, deleting, modifying, and querying the cultural element information, and the information in the database can be constantly updated and modified.

(2) Graphic elements extraction.

The graphic element extraction module is used to extract the basic elements for each graph in the Chinese ethnic apparel database, and then store the basic elements in the basic elements database. The graphic element extraction module is used to obtain all the basic elements contained in the Chinese ethnic apparel database and indirectly serves the association rule mining module.

(3) Basic elements database.

The basic element database stores the basic elements extracted from the graphics extraction module, and the information of each basic element includes theme, shape, dynasty, horizontal symmetry, central symmetry, and vertical symmetry. The basic element database directly serves the association rule mining module and records the information of all the basic elements.

(4) Association rule mining.

The association rule mining module is the core module of the GDAssister, which not only mining the association between the basic elements but also mining the association rules between the basic elements and the skeleton of the graph. The module provides inspiration, and without it, the recommendation results will be completely dependent on the user's input. Therefore, we introduce relevance rules to describe the relationship between elements and elements, we can recommend the material associated with the input material and generate new ideas.

(5) Automatic layout.

The automatic layout module receives the input from the user, then uses association rules to generate similar basic elements and skeletons, and then combines different elements and skeletons to generate multiple graphs, which are displayed to the user at the front end.

(6) Style transform.

The style transform module is based on style transfer technology, and many color schemes are designed in advance according to the common needs of designers, such as Edward Monk's famous painting "Scream", Van Gogh's famous painting "Starry Sky" and so on, as shown in the following picture. The module receives the design drawings from the layout module and renders different color styles for designers.

4   Implementation of GDAssister

4.1   Graphic Elements Extraction

The graphic element extraction module uses content-based image retrieval technology to complete the extraction of elements. Content-based image retrieval technology is used to analyze and classify the input image, extract its color, shape, texture, outline, and spatial location and other features, establish a feature index and store it in the feature database. During retrieval, the user submits the source image of the query, sets the query conditions through the user interface, and can be represented by one or more combinations of features, and then extract the desired associated image from the image database and feed it back to the user in the order of similarity from large to small. According to their own restrictions, designers can choose whether to modify the query conditions and continue to query in order to achieve satisfactory query results.

The core of the module is the image feature database. Features are extracted from the image itself and used to calculate the similarity between images. The module's framework mainly include the following basic functional modules: retrieval method setting, retrieval results browsing, database management, and maintenance, as shown in Fig. 2.

The module is implemented by VGGNet16 [22], VLAD(Vector Of Locally Aggregated Descriptors) algorithm [15], and PQ(Product Quantization) quantization algorithm [12]. VGGNet16 is responsible for extracting image features, VLAD improves the search efficiency and accuracy by reducing the dimension of

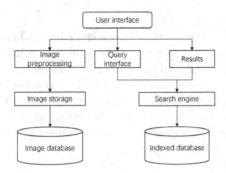

Fig. 2. Structure of graphic elements extraction module.

the image features, PQ quantization is used to accelerate the distance calculation between the feature vectors obtained by the VLAD algorithm.

4.2   Association Rule Mining

Association rule mining is to generate validity rules. These rules' support and confidence benchmark are greater than the minimum benchmark set by the user. Mining association rules is usually divided into two steps: firstly, finding out that all the occurrence frequencies are bigger than the predefined minimum support degree, secondly, generating rules by utilizing frequent itemsets.

General cultural graphs are composed of a certain number of circular units, which refer as elements. A graph generally consists of one or more element units. The structure of elements in the graph is called skeleton. An graph element unit is the smallest design unit in a graph, and contains the following descriptive information.

- The identification id of the element. The identification id consists of two parts. One is the graph id of the element, the element of the same graph has the same graph id, and the other is the element id, which is the id of the element in the element database.
- The content information about the element, including the theme, age, way of expression, etc.
- Structural information. The structure information of graphic unit describes the shape, symmetry information, and classification information. The shape not only includes circle, triangle, rectangle, fan, and other basic shapes, but also complex shapes such as heart shape, instrument shape, cloud color shape, and so on. Symmetry include horizontal symmetry, vertical symmetry, and central scale.

The skeleton represents the arrangement rule of the graph unit in the graph, and it is the minimum repeatable structure in the graph. Its function is to determine the arrangement and expression of the elements. The skeleton has two kinds of information, one is the location information about the elements,

Algorithm 1. Product quantization algorithm in graphic element extraction

```
 1: function SPATIAL_SEGMENTATION(d,nsq)
 2:     int ds=d/nsq
 3:     for i = 0; i<nsq; i + + do
 4:         for j = 0; j<ds; j + + do
 5:             vs.pushback(vtrain.row(i * ds + j))
 6:         end for
 7:     end for
 8: end function
 9:
10: function QUANTIFICATION(Mat mat1, Mat mat2)
11:     int d=mat1.cols
12:     float sum_d=0.0
13:     for i = 0; i<d; i + + do
14:         float pp=mat1.at < float > (0, i) − mat2.at < float > (0, i)
15:         sum_d=sum_d+pp * pp
16:     end for
17:     return sum_d
18: end function
19:
20: function DISTANCE_CALCULATION(Mat D, Mat x, Mat dis)
21:     dis=Mat::zero(1, x.rows, CV_3 2FC1)
22:     for i = 0; i<x.rows; i + + do
23:         float distmp=0
24:         for j = 0; j<D.cols; j + + do
25:             distmp=distmp + D.at < float > (x.at < int > (i, j), j)
26:         end for
27:         dis.at < float > (0, i)=distmp
28:     end for
29:     return dis
30: end function
```

which is described by coordinating parameters and defines positions need to be placed in a canvas, and the other is the nature of the position itself, such as the number, size, flipping mode and spin angle of the element in this position. Designed graphics can be created by combining the defined skeleton with the selected elements.

The most representative algorithms for association rule mining are Apriori algorithm and FP-Growth algorithm. However, These traditional association rule mining algorithms will produce redundant frequent item sets of worthless association rules. Although FP-Growth algorithm has its own advantages over Apriori algorithm, the process of rule mining using FP-Growth algorithm in most databases requires a long time and a large amount of main memory. Therefore, we propose an improved FP-Growth algorithm, named IFP-Growth algorithm (Improved FP-Growth). We use IFP-Growth algorithm to mine frequent itemsets in Chinese-style database, and finally get the association rules between elements and skeletons.

IFP-Growth's key points are as follows.

- Creating RFP-tree and CFP-tree. RFP-tree is responsible for rare data, and CFP-tree is responsible for commom data. RFP-tree is expected to be much smaller than CFP-tree.
- Using the mining process of FP-Growth algorithm to mine RFP-tree, and obtaining frequent itemsets of rare data.
- The flag item is introduced in the process of establishing CFP-tree, which is convenient to reduce the cyclic recursive process of mining frequent itemsets, so as to improve the efficiency of the algorithm.

IFP-Growth's process is as follows.

- If itemset contains rare items. Fisrtly, extracting conditional pattern library. Starting with each frequent item in the header table, getting the conditional pattern base of each item, and the prefix path that needs to be deleted has been marked with a Boolean flag during the establishment of the CFPtree, which will greatly reduce the next step of recursive calls. Secondly, creating a FP-tree that belongs to a frequent item. Creating a conditional FP-tree that belongs to that frequent item for each frequent item. In the process of using cyclic recursion to obtain the prefix path of frequent items, when the Boolean flag is judged to be zero, we immediately end the prefix path search of the node, go back to the previous node to continue the process of cyclic recursion, and finally get all the prefix paths of the item. Using the above process to build the FP-tree that belongs to this item. Thirdly, inputting minimum support to get frequent itemsets. According to the FP-tree of each frequent item in the second step, the frequent itemset satisfying the condition can be obtained according to the process of FP-Growth algorithm.
- If itemset does not contain rare items. Use the original FP-Growth process to mine RFP-tree to obtain frequent itemsets.

We use Apriori, FP-Growth and IFP-Growth to mine association rules. Under different minimum support, Fig. 3 show that in the process of mining association rules in the silk cultural relic database, the Apriori has the longest execution time, because Apriori needs to scan the whole database frequently. Compared with Apriori algorithm, FP-Growth algorithm has been greatly improved the mining efficiency. However, when the database is large, the FP-tree will also be very large. IFP-Growth reduces the running time of the whole algorithm by reducing the recursive calls in the process of finding frequent itemsets.

At present, the Chinese ethnic apparel database contains more than 5000 graphs. It concludes 1000 elements approximatly and more than 300 kinds of skeletal structures. It is meaningless for these graphic elements to participate in association data mining as a single individual, and it is necessary to properly classify elements and skeletons. Similar elements or skeletons are grouped into a large category, and association rules are used to describe the relationship between different categories. In order to maintain the number of reasonable individuals in each category, the criteria for classification should be neither too detailed nor too broad.

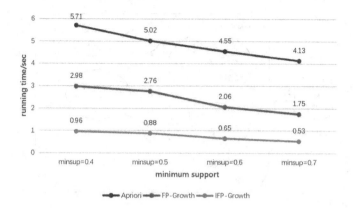

Fig. 3. The runing time of Apriori, FP-Growth, and IFP-Growth under different minimum support.

The id here is not the same id that in the actual database, which only marks the categories in the process of association rule mining. In fact, many element id will have the same mapping to an element id. A graph can be defined as a transaction, such as $(Transaction1, P1, P2, S1)$, $(Transaction2, P1, P3, P5, S2)$, etc. Using Pi to represent the id of the element. Use Si to represent the id of the skeleton. These transactions use the data structure of list as the data set to input into the process of IFP-Growth mining, and the mining results are the following two rules.

$$[(P1, P2 \rightarrow P3), \gamma] \tag{1}$$

$$[(P1, P2 \rightarrow P3), \gamma] \tag{2}$$

The first rule describes the relationship between elements, especially the relationship between the main graph and the auxiliary graph, which is derived from the main graph or the combination of the main graph and the auxiliary graph. The second rule describes the relationship between the element and the skeleton structure. The rule deduces the skeleton structure according to the main graph and auxiliary graph. We refer to gamma as the confidence of the normalized rules, which indicating the strength of rules. The generated rules are stored according to key-value pairs in the form of Map <String, Integer>, and the required association rules are sorted according to the gamma from high to low. At the end of the association mining process, these association rules will be stored in the rule database. Figure 4 shows the process of association rules mining.

4.3   Automatic Layout

The automatic layout module provides design tools for designers at the portal web. This module includes layer design and pattern editing. Pattern editing includes the processing of pattern size, position, angle, arrangement and so on.

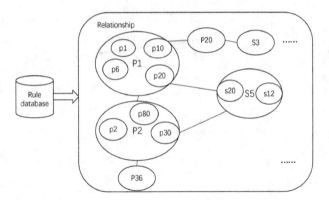

Fig. 4. The process of mining association rules using IFP-Growth algorithm.

The graphic elements come from the database, the design scheme comes from the output of the association rule mining module, and the designer adjusts and optimizes the design drawing on the basis of the recommended scheme. The design drawing is sent to the style transform module after the format conversion is carried out on the background server. Figure 5 shows the result of automatic layout.

Fig. 5. Automatic layout presentation

4.4 Style Transform

The style transform module is implemented by color-based style transfer, which refers to the use of algorithms to learn the style of famous paintings, and then apply this style to another picture. The module contains two network models, one is the loss network, the other is the generation network. The dataset uses the coco dataset. Figure 6 shows the training model with shouting style and an case of style conversion.

The loss network is implemented by the image recognition model VGGNet. The convolution layer in front of VGGNet extracts "features" from the image, while the later full connection layer converts the "features" of the image into category probability. Among them, the features extracted from the shallow layer are often relatively simple, and the features extracted from the deep layers are

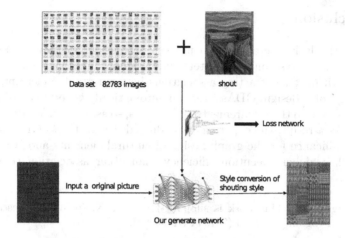

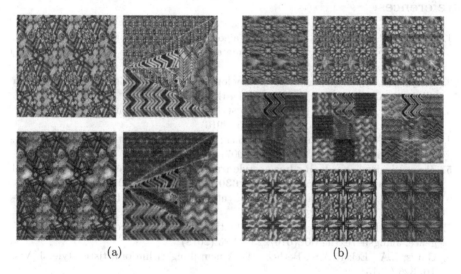

Fig. 6. The training process of shouting style model and an case of style conversion.

(a)   (b)

Fig. 7. Style transform presentation.

often more complex. The process of VGGNet is to input images, extract features, output image categories, image style transfer is just the opposite, input features, output the corresponding pictures, as shown in the following figure. Specifically, the style transfer uses the middle feature of the convolution layer to restore the original image corresponding to this feature.

Loss network is used to define content loss and style loss, the total loss is composed of content loss and style loss, the degree of content loss and style loss is controlled by setting different weights, and the total loss is used to train the image generation network. Each generation network represents a style. Figure 7 shows the result of style transform.

5 Conclusion

In view of the lack of creativity in modern art design, we propose an assisting graphic design system(GDAssister) and design an improved association rule mining algorithm for extracting associations between graphic elements. In the process of graphic design, GDAssister can automatically generate corresponding graphs according to the requirements of designers, so as to provide inspiration for designers. As a result, the experiment conducted demonstrates GDAssister has positive significance for the graph design of cultural elements and IFP-Growth algorithm has higher execution efficiency than other association rule mining algorithms.

Acknowledgement. This work is supported by The National Key Research and Development Program of China (2019YFB1405702).

References

1. Agrawal, R., Srikant, R., et al.: Fast algorithms for mining association rules. In: Proceedings of 20th International Conference on Very Large Data Bases, VLDB, vol. 1215, pp. 487–499 (1994)
2. Canjun, H.: Development of key technologies of creative design expert system for silk cultural relics based on Web. Ph.D. thesis (2018)
3. Chen, Y.L., Hsu, C.T.: Towards deep style transfer: a content-aware perspective. In: British Machine Vision Conference (2016)
4. Liu de and Wang de Huang: Application of fuzzy reasoning in textile expert system. Wool Spinning Technol. **02**, 52–54 (2007)
5. Zhou, W., Dou, Y., Ji, R.: Image style transfer technique based on convolutional neural network. Mod. Comput. (Pro) **630**(30), 49–53+62 (2018)
6. Efros, A.A.: Image quilting for texture synthesis and transfer. In: Computer Graphics (2001)
7. Fajun, Y.: Development measures of textile technology based on artificial intelligence. Jiangsu Textile **000**(010), 23–24,26 (2018)
8. Gatys, L.A., Ecker, A.S., Bethge, M.: A neural algorithm of artistic style. J. Vis. **16**, 326 (2015)
9. Chen, G.: Research on image Style Transfer Method based on generative adversarial network. Ph.D. thesis, Beijing University of Posts and Telecommunications (2019)
10. Peng, G.: Color matching expert system based on silk cultural relics. Ph.D. thesis (2017)
11. Han, J., Pei, J.: Mining frequent patterns by pattern-growth: methodology and implications. ACM SIGKDD Explor. Newslett. **2**(2), 14–20 (2000)
12. Jegou, H., Douze, M., Schmid, C.: Product quantization for nearest neighbor search. IEEE Trans. Pattern Anal. Mach. Intell. **33**(1), 117–128 (2010)
13. Ai, J.: Design and implementation of painting art style rendering system based on deep learning. Ph.D. thesis (2020)
14. Zhang Lingli, L., Dongming, P.Y., Zhengsan, J.: Composition knowledge generation model based on comprehensive reasoning. J. Comput. Aid. Des. Graph. **05**, 384–389 (2000)

15. Mironica, I., Duta, I., Ionescu, B., Sebe, N.: Beyond bag-of-words: fast video classi-fication with fisher kernel vector of locally aggregated descriptors. In: IEEE International Conference on Multimedia and Expo (2015)

16. Müller, H., Michoux, N., Bandon, D., Geissbuhler, A.: A review of content-based image retrieval systems in medical applications–clinical benefits and future directions. Int. J. Med. Inf. **78**(9), 638–638 (2004)

17. Li, S.: Application of image ink style rendering based on deep learning. Ph.D. thesis (2017)

18. Sun, D., Youdong, D., Yun, Q.: Application of style transfer algorithm based on artificial intelligence in animation special effects design. Decoration **000**(001), 104–107 (2018)

19. Valgren, C.: SIFT, SURF and seasons: long-term outdoor localization using local features. In: Proceedings of ECMR (2007)

20. Fu, Y.: Research on creative design method of color matching scheme based on silk cultural relics. Ph.D. thesis

21. Zhang, L., Ji, Y., Lin, X., Liu, C.: Style transfer for anime sketches with enhanced residual u-net and auxiliary classifier GAN. In: 2017 4th IAPR Asian Conference on Pattern Recognition (ACPR), pp. 506–511. IEEE (2017)

22. Zhu, X., Yuan, J., Xiao, Y., Zheng, Y., Qin, Z.: Stroke classification for sketch segmentation by fine-tuning a developmental VGGNet16. Multimed. Tools Appl. **79**(6), 33891–33906 (2020)

Matrix Computation

Late Fusion Multi-view Clustering with Learned Consensus Similarity Matrix

Dan Liu, Weixuan Liang$^{(\boxtimes)}$, Hongya Zhang, Wentao Zhao, and Kaifan Hu

National University of Defense Technology, Changsha 410073, Hunan, China
weixuanliang@nudt.edu.cn

Abstract. Multiple kernel algorithms in a late fusion manner have been widely used because of its excellent performance and high efficiency in multi-view clustering (MVC). The existing MVC algorithms via late fusion obtain a consensus clustering indicator matrix through the linear combination of the base clustering indicator matrix. As a result, the optimal consensus indicator matrix's searching space reduces, and the clustering effect is limited. To learn more information from the base clustering indicator matrices, we construct a consensus similarity matrix as the input of the spectral clustering algorithm. Furthermore, we design an effective iterative algorithm to solve the new resultant optimization problem. Extensive experiments on 11 multi-view benchmark datasets demonstrate the effectiveness and efficiency of the proposed algorithm.

Keywords: Late fusion · Kernel k-means · Similarity matrix

1 Introduction

Many datasets in the real world are naturally comprised of different representations or views. Observing that these multiple representations often provide compatible and complementary information, it becomes natural to integrate them to obtain better performance rather than rely on a single view. Because data represented by different views that can be heterogeneous and biased, how to fully utilize the multi-view information to obtain a consensus representation for clustering is a challenging problem in multi-view clustering (MVC). In recent years, various kinds of algorithms are developed for a better clustering performance or higher efficiency, such as multi-view subspace clustering [4,17,18], multi-view spectral clustering [7,9,19], multiple kernel k-means (MKKM) [5,6,8,11,12]. Among them, MKKM has been widely used due to its excellent clustering performance and strong interpretability.

According to the stage of multi-view data fusion, the existing literature of MVC can be roughly dividing into two categories. **The first category**, which can be described as early fusion, fuses the information of all views before clustering. Huang et al. [6] optimize an optimal kernel matrix as a linear combination of a set of pre-specified kernels for clustering. To reduce the redundancy of the predefined kernels, Liu et al. [8] propose an MKKM algorithm with matrix-induced

© Springer Nature Singapore Pte Ltd. 2021
K. He et al. (Eds.): NCTCS 2020, CCIS 1352, pp. 73–86, 2021.
https://doi.org/10.1007/978-981-16-1877-2_6

regularization. To enhance the clustering performance, a local kernel alignment criterion is adopted for MKKM in [8]. The proposed algorithm in [12], allows the optimal kernel to reside in the neighbor of the combinational kernels, which increases the searching the space of the optimal kernel.

Although the afore-mentioned early fusion algorithms have greatly improved the clustering accuracy of MKKM from different aspects, they have two drawbacks that should not be neglected. One is the intensive computational complexity, i.e., usually $\mathcal{O}(n^3)$ per iteration (n is the number of samples). The other is that the over-complicated optimization processes of these methods, which increases the risk of being trapped into bad local minimums, leading to unsatisfying clustering performance.

The second category, which is termed as late fusion, maximizes the alignment between the consensus clustering indicator matrix and the weighted base clustering indicator matrices with orthogonal transformation, where each clustering indicator matrix is generated by performing clustering on each single view [10,16]. The proposed method in this paper belongs to the second category. This kind of algorithms have improved efficiency and clustering performance. However, these methods simplify the searching space of the optimal clustering indicator matrix, resulting in a poor clustering performance.

In this paper, to extract more information from the base clustering indicator matrix, we propose a new late fusion framework. We assume that each base clustering indicator matrix is a feature representation after the sample is mapped from the original space to the feature space. Compared with the original features, it involves the nonlinear relation between data [15] and also can filter out some noise of the kernel matrix. Therefore, the similarity matrix constructed by the base clustering indicator matrix is better than the original kernel matrix for clustering.

We term the proposed method as late fusion multi-view clustering with learned consensus similarity matrix (LF-MVCS). LF-MVCS jointly optimizes the combination coefficient of the base clustering indicator matrices and the consensus similarity matrix. And this consensus similarity matrix is then used as the input of spectral clustering [14] to obtain the final result. Besides, we design an effective algorithm to solve the resulting optimization problem and analyze its computational complexity and convergence. Extensive experiments on 11 multi-view benchmark datasets are conducted to evaluate the effectiveness and efficiency of our proposed method.

The contributions of this paper are summarized as follows,

- The proposed LF-MVSC integrates the information of all views in a late fusion manner. It optimizes the combination coefficient of each base clustering indicator matrix and consensus similarity matrix. As far as we know, this is the first time to obtain a unified form that can better reflect the similarity between samples via late fusion.
- The proposed optimization formulation has no parameter to tune and avoids the time consuming caused by searching optimal parameters. We design an alternate optimization algorithm with proved convergence to efficiently tackle

the resultant problem. Our algorithm avoids complex computation because of fewer optimization variables so that the objective function can converge within fewer iterations.

2 Related Work

2.1 Multiple Kernel K-Means

Let $\{\mathbf{x}_i\}_{i=1}^n \subseteq \mathcal{X}$ be n independent and identically distributed samples, and $\phi(\cdot) : \mathbf{x} \in \mathcal{X} \to \mathcal{H}$ be a feature mapping function, which transfers $\mathbf{x}$ into a reproducing kernel Hilbert space $\mathcal{H}$. The objective of kernel k-means clustering aims to minimize a sum-of-squared loss function over the cluster assignment matrix $\mathbf{Z} \in \{0,1\}^{n \times k}$, which can be formulated as follows,

$$\min_{\mathbf{Z} \in \{0,1\}^{n \times k}} \sum_{i=1,j=1}^{n,k} Z_{ij} \|\phi(\mathbf{x}_i) - \boldsymbol{\mu}_j\|_2^2 \; s.t. \sum_{c=1}^{k} Z_{ij} = 1, \tag{1}$$

where $\boldsymbol{\mu}_j = \frac{1}{n_j} \sum_{i=1}^n Z_{ij} \phi(\mathbf{x}_i)$ is the centroid of cluster j $(1 \leq j \leq k)$, and $n_j = \sum_{i=1}^n Z_{ij}$ is the number of samples in cluster $\mathcal{C}_j$.

To optimize Eq. (1), we can transform it as follows,

$$\min_{\mathbf{Z}} \mathrm{Tr}(\mathbf{K}) - \mathrm{Tr}(\mathbf{A}^\top \mathbf{K} \mathbf{A}) \; s.t. \; \mathbf{A}^\top \mathbf{A} = \mathbf{I}_k, \tag{2}$$

where $\mathbf{K}$ is a kernel matrix with $K_{ij} = \phi(\mathbf{x}_i)^\top \phi(\mathbf{x}_j)$, and $\mathbf{A} \in \mathbb{R}^{n \times k}$ is a discrete clustering indicator matrix as

$$A_{ij} = \begin{cases} \frac{1}{\sqrt{n_j}}, & \text{if } \mathbf{x}_i \in \mathcal{C}_j \\ 0, & \text{if } \mathbf{x}_i \notin \mathcal{C}_j, \end{cases} \tag{3}$$

The discrete constraint of $\mathbf{A}$ makes the optimization problem in Eq. (2) NP-hard to solve. Dhillon et al. [2] relax the discrete constraint and let $\mathbf{A}$ take arbitrary real values subject to orthogonality constraints. We denote the relaxed $\mathbf{A}$ as $\mathbf{H}$. Thus, Eq. (2) can be rewritten as:

$$\max_{\mathbf{H}} \mathrm{Tr}(\mathbf{H}^\top \mathbf{K} \mathbf{H}) \; s.t. \; \mathbf{H}^\top \mathbf{H} = \mathbf{I}_k, \tag{4}$$

where $\mathbf{I}_k$ is an identity matrix with size $k \times k$. Finally, the optimal $\mathbf{H}$ can be obtained for Eq. (4) by taking the k eigenvectors that correspond to the k largest eigenvalues of $\mathbf{K}$. The Lloyd's algorithm is usually performed on $\mathbf{H}$ for the clustering labels.

In multiple kernel k-means (MKKM), all samples have multiple feature representations via a group of feature mappings $\{\phi(\cdot)\}_{p=1}^m$. Specifically, each sample can be represented as $\phi_\alpha(\mathbf{x}) = [\alpha_1 \phi_1(\mathbf{x})^\top, ..., \alpha_m \phi_m(\mathbf{x})^\top]^\top$, where $\boldsymbol{\alpha} = [\alpha_1, ..., \alpha_m]^\top$ are the coefficients of base kernels. Assume that the optimal kernel matrix can be represented as $\mathbf{K}_\alpha = \sum_{p=1}^m \alpha_p^2 \mathbf{K}_p$, we obtain the optimization objective of MKKM as follows,

$$\min_{\mathbf{H},\alpha} \mathrm{Tr}(\mathbf{K}_\alpha (\mathbf{I}_n - \mathbf{H}\mathbf{H}^\top)) \tag{5}$$
$$\text{s.t. } \mathbf{H} \in \mathbb{R}^{n\times k} , \ \mathbf{H}^\top \mathbf{H} = \mathbf{I}_k , \ \alpha^\top \mathbf{1} = 1 , \ \alpha \geq 0.$$

A two-step iterative algorithm can be used to optimize the Eq. (5). **i) Optimize α with fixed H.** Eq. (5) is be equivalent to:

$$\min_\alpha \sum_{p=1}^m \alpha_p^2 \delta_p , \text{ s.t. } \alpha^\top \mathbf{1} = 1 , \ \alpha \geq 0. \tag{6}$$

where $\delta_p = \mathrm{Tr}(\mathbf{K}_p (\mathbf{I}_n - \mathbf{H}\mathbf{H}^\top))$. According to the Cauchy-Schwartz's inequality, we can obtain the closed-form solution of Eq. (6) as $\alpha_p = \frac{1/\sqrt{\delta_p}}{\sum_{p=1}^m 1/\sqrt{\delta_p}}(\forall p \in [m])$. **ii) Optimize H with fixed α.** With the kernel coefficients α fixed, the **H** can be optimized by taking the k eigenvectors that correspond to the k largest eigenvalues of $\mathbf{K}_\alpha$.

2.2   Multi-view Clustering via Late Fusion Alignment Maximization

The multi-view clustering via late fusion alignment maximization [16] (MVC-LFA) assumes that the clustering indicator matrix obtained from each single view is $\mathbf{H}_p(p \in [m])$, and a set of rotation matrices $\{\mathbf{W}_p\}_{p=1}^m$ are used to maximize the alignment of $\mathbf{H}_p(p \in [m])$ and the optimal clustering indicator matrix $\mathbf{H}^*$. The objective function is as follows,

$$\max_{\mathbf{H}^*,\{\mathbf{W}_p\}_{p=1}^m,\mu} \mathrm{Tr}(\mathbf{H}^{*\top}\mathbf{R}) + \lambda \mathrm{Tr}(\mathbf{H}^{*\top}\mathbf{F})$$
$$\text{s.t. } \mathbf{H}^* \in \mathbb{R}^{n\times k} , \ \mathbf{H}^{*\top}\mathbf{H}^* = \mathbf{I}_k , \ \mathbf{W}^\top \mathbf{W} = \mathbf{I}_k, \tag{7}$$
$$\sum_{p=1}^m \mu_p^2 = 1 , \ \mu_p \geq 0 , \ \mathbf{R} = \sum_{p=1}^m \mu_p \mathbf{H}_p \mathbf{W}_p,$$

where $\mathbf{F}$ denotes the average clustering indicator matrix and λ is a trade-off parameter. The latter $\mathrm{Tr}(\mathbf{H}^{*\top}\mathbf{F})$ is a regularization on optimal clustering indicator matrix to prevent $\mathbf{H}^*$ from being too far way from prior average partition. This formula assumes that the optimal clustering indicator matrix $\mathbf{H}^*$ is a linear combination of the base clustering indicator matrix $\mathbf{H}_p(p \in [m])$. One can refer to [16] for the optimization of Eq. (7).

MVC-LFA can reduce the complexity in each iteration [16], but it oversimplifies the searching space of the optimal clustering indicator matrix, which limits the clustering performance. To solve this problem, in Sect. 3, we first construct a consensus similarity matrix through the base clustering indicator matrices and then use the spectral clustering algorithm to obtain the final clustering result (Fig. 1).

3 Late Fusion Multi-view Clustering with Learned Consensus Similarity Matrix

Learning Algorithm

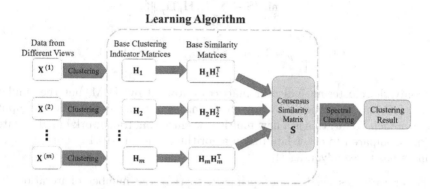

Fig. 1. Model pipeline for the proposed *Late Fusion Multi-view Clustering with Learned Consensus Similarity Matrix* (LF-MVCS). Firstly, we obtain the base clustering indicator matrix independently from each view by kernel k-means clustering. Then, we jointly optimize the combination coefficient of the base clustering indicator matrices and the consensus similarity matrix. Finally, this consensus similarity matrix is used as the input of spectral clustering to obtain the final result.

3.1 Proposed Method

By taking the eigenvectors corresponding to the first k largest eigenvalues of $\{\mathbf{K}_p\}_{p=1}^m$, we obtain the clustering indicator matrices $\{\mathbf{H}_p\}_{p=1}^m \subseteq \mathbb{R}^{n \times k}$. Each row of $\mathbf{H}_p$ can be regarded as a feature representation of each sample in the k dimension space. To construct the consensus similarity matrix $\mathbf{S}$ we need, we need to construct the similarity matrix of each view. After normalizing each row of $\mathbf{H}_p$, i.e., divided by the 2-norm of the row, the 2-norm of each row of the processed $\mathbf{H}_p$ becomes 1. We treat $\mathbf{S}_p = \mathbf{H}_p\mathbf{H}_p^\top \in \mathbb{R}^{n \times n}$ as the similarity matrix of the p-th view.

Based on the assumption that the optimal consensus similarity matrix resides in the neighbor of a linear combination of all the similarity matrices, our idea can be fulfilled as follows,

$$\min_{\mathbf{S},\boldsymbol{\alpha}} \|\mathbf{S} - \sum_{p=1}^m \alpha_p \mathbf{H}_p\mathbf{H}_p^\top\|_{\mathrm{F}}^2, \text{ s.t. } \sum_{p=1}^m \alpha_p = 1, \ \alpha_p \geq 0. \tag{8}$$

The solution of the above formula is very sensitive to initialization. Once $\boldsymbol{\alpha}$ is fixed, $\mathbf{S}$ becomes a fixed value, which may not be optimal, resulting in an unsatisfactory solution. Hence, we have to impose some mild constraints on $\mathbf{S}$. We assume that: *i) The degree of each sample in $\mathbf{S}$ is greater than or equal to 0,* *i.e.*, $S_{ij} \geq 0$; *ii) The degree of each sample in* $\mathbf{S}$ *is equal to* 1, *i.e.*, $\mathbf{S}^\top \mathbf{1} = \mathbf{1}$. In this way, we can obtain the following formula:

$$\min_{\mathbf{S},\alpha} \|\mathbf{S} - \sum_{p=1}^{m} \alpha_p \mathbf{H}_p \mathbf{H}_p^\top \|_{\mathrm{F}}^2,$$

$$\text{s.t.} \sum_{p=1}^{m} \alpha_p = 1 , \ \alpha_p \geq 0 , \ S_{ij} \geq 0 , \ \mathbf{S}^\top \mathbf{1} = \mathbf{1},$$

(9)

Not only the clustering indicator matrices obtained by KKM, but the similar basic partitions constructed by any clustering algorithms also can be the input of our model. So far, this is a new multi-view clustering framework based on late fusion. Compared to other late fusion algorithms, such as [10,16], our objective formula has three advantages:

- Fewer variables need to be optimized so that the number of iterations is reduced, and the algorithm converges quickly.
- Other late fusion algorithms believe that the optimal clustering indicator matrix is a neighbor of the average clustering indicator matrix, which is a heuristic assumption and may not work in some cases. Our assumptions of the Eq. (9) are more reasonable, and will be effective in any multi-view clustering tasks.
- Our objective formula has no regularization term. Compare with most multi-view algorithms, our algorithm avoids the extra computational overhead caused by searching for optimal hyperparameters.

3.2 Alternate Optimization

In the following, we design an efficient algorithm to solve the optimization problem in Eq. (9). In specific, we design a two-step algorithm to solve it alternately:

Optimization α with Fixed S
With $\mathbf{S}$ being fixed, the optimization problem in Eq. (9) is equivalent to the optimization problem as follows,

$$\min_{\alpha} \frac{1}{2} \alpha^\top \mathbf{M} \alpha + \mathbf{f}^\top \alpha, \ \text{s.t.} \sum_{p=1}^{m} \alpha_p = 1 , \ \alpha_p \geq 0,$$

(10)

where $\mathbf{M} \in \mathbb{R}^{m \times m}$ with $M_{pq} = 2\mathrm{Tr}(\mathbf{H}_p \mathbf{H}_p^\top \mathbf{H}_q \mathbf{H}_q^\top)$ and $\mathbf{f} \in \mathbb{R}^m$ with $f_p = -2\mathrm{Tr}(\mathbf{S}^\top \mathbf{H}_p \mathbf{H}_p^\top)$. $\mathbf{M}$ is a nonnegative definite matrix, so the optimization in Eq. (10) w.r.t α is a quadratic programming problem with linear constraint.

Optimization S with Fixed α
With α being fixed, the optimization problem in Eq. (9) w.r.t $\mathbf{S}$ can be rewritten as,

$$\min_{\mathbf{S}} \|\mathbf{S}\|_{\mathrm{F}}^2 - \mathrm{Tr}(\mathbf{S}^\top \mathbf{G})$$

$$\text{s.t.} \ S_{ij} \geq 0 , \ \mathbf{S}^\top \mathbf{1} = \mathbf{1},$$

(11)

where $\mathbf{G} = 2 * \sum_{p=1}^{m} \alpha_p \mathbf{H}_p \mathbf{H}_p^\top$. Note that the problem in Eq. (11) is independent between j; we can optimize each column of $\mathbf{S}$ as:

$$\min_{\mathbf{s}_j} \mathbf{s}_j^\top \mathbf{s}_j - \mathbf{s}_j^\top \mathbf{g}_j$$
$$\text{s.t. } s_{ij} \geq 0, \ \mathbf{s}_j^\top \mathbf{1} = 1, \tag{12}$$

where $\mathbf{s}_j$ is the j-th column of $\mathbf{S}$ and $\mathbf{g}_j$ is the j-th column of $\mathbf{G}$. Equation (12) is equivalent to the following function,

$$\min_{\mathbf{s}_j} \frac{1}{2} \|\mathbf{s}_j - \frac{\mathbf{g}_j}{2}\|_2^2$$
$$\text{s.t. } s_{ij} \geq 0, \ \mathbf{s}_j^\top \mathbf{1} = 1, \tag{13}$$

Equation (13) is a problem on the simplex space and its Lagrangian function is

$$\mathcal{L}(\mathbf{s}_j, \gamma, \boldsymbol{\tau}) = \frac{1}{2} \|\mathbf{s}_j - \frac{\mathbf{g}_j}{2}\|_2^2 + \gamma(\mathbf{s}_j^\top \mathbf{1} - 1) - \boldsymbol{\tau}^\top \mathbf{s}_j \tag{14}$$

where γ and $\boldsymbol{\tau}$ are the Lagrangian multipliers.

According to its Karush-Kuhn-Tucker condition [1], we can obtain the optimal solution $\mathbf{s}^*$ as

$$\mathbf{s}_j^* = (\frac{\mathbf{g}_j}{2} - \gamma^* \mathbf{1})_+ \tag{15}$$

where $(\mathbf{x})_+$ is a signal function, which sets the elements of $\mathbf{x}$ less than 0 to 0, and $\gamma^* \in \mathbb{R}$ makes $\mathbf{1}^\top \mathbf{s}_j^* = 1$.

The detailed algorithm is summarized in Algorithm 1.

Algorithm 1. Late Fusion Multi-view Clustering with Learned Consensus Similarity Matrix

---

Input: $\{\mathbf{H}_p\}_{p=1}^m$ and ϵ.
Output: S.
1: Initialize $\boldsymbol{\alpha}$ as $\mathbf{1}_m/m$ and t as 1.
2: **while** not converge **do**
3:     **for** $j \in \{1, 2, ..., n\}$ **do**
4:         Update $\mathbf{s}_j$, by solving (13) with fixed $\boldsymbol{\alpha}$.
5:     **end for**
6:     $\mathbf{S} = \frac{\mathbf{S} + \mathbf{S}^\top}{2}$.
7:     Update $\boldsymbol{\alpha}$, by solving (10) with fixed $\mathbf{S}$.
8:     $t = t + 1$.
9: **end while**$\|\boldsymbol{\alpha}^{(t)} - \boldsymbol{\alpha}^{(t-1)}\|_2^2 \leq \epsilon$

3.3 Algorithmic Discussion

Computational Complexity. Obtaining the base clustering indicator matrices needs to calculate the eigenvectors corresponding to the k largest eigenvalues of the m kernel matrices, which leads to $\mathcal{O}(kmn^2)$ complexity. The proposed iterative algorithm needs to optimize two variables $\boldsymbol{\alpha}$ and $\mathbf{S}$. In each iteration, the optimization of $\boldsymbol{\alpha}$ by Eq. (10) requires solving a standard quadratic programming problem with linear constraint whose complexity is $\mathcal{O}(m^3)$. Updating each $\mathbf{s}_j$ by Eq. (13) requires $\mathcal{O}(n)$ time. The complexity of obtaining $\mathbf{S}$ per iteration is $\mathcal{O}(n^2)$, due to n times are needed to calculate each $\mathbf{s}_j$. Overall, the total complexity of Algorithm 1 is $\mathcal{O}((m^3 + n^2)t)$, where t is the number of iterations. Finally, performing the standard spectral clustering algorithm on $\mathbf{S}$ takes $\mathcal{O}(n^3)$ time.

Convergence. In each of the optimization iteration of our proposed algorithm, the optimizations of $\boldsymbol{\alpha}$ and $\mathbf{S}$ monotonically decrease the value of the objective function (9). At the same time, the objective is lower-bounded by zero. As a result, our algorithm is guaranteed to converge to a local optimum of problem Eq. (9).

4   Experiments

4.1   Datasets and Experimental Settings

Several benchmark datasets are used to demonstrate the effectiveness of the proposed method, including *Flower17*[1], *Flower102*[2], *PFold*[3], *Digit*[4], *CCV*[5] and *Cal*[6]. Six sub-datasets which are constructed by selecting the first 5, 10, 15, 20, 25 and 30 classes from Cal data, are also used in our experiment. The detailed information of these datasets is listed in Table 1. From this table, we observe that the number of samples, views and the number of categories of these datasets range from 510 to 8189, 3 to 48, and 5 to 102, respectively.

Table 1. Benchmark datasets

Datasets	Samples	Kernels	Clusters
Flower17	1360	7	17
Flower102	8189	4	102
PFold	694	12	27
Digit	2000	3	10
CCV	6773	3	20
Cal-5	510	48	5
Cal-10	1020	48	10
Cal-15	1530	48	15
Cal-20	2040	48	20
Cal-25	2550	48	25
Cal-30	3060	48	30

[1] www.robots.ox.ac.uk/~vgg/data/flowers/17/.
[2] www.robots.ox.ac.uk/~vgg/data/flowers/102/.
[3] mkl.ucsd.edu/dataset/protein-fold-prediction.
[4] http://ss.sysu.edu.cn/py/.
[5] www.ee.columbia.edu/ln/dvmm/CCV/.
[6] www.vision.caltech.edu/Image_Datasets/Caltech101/.

Table 2. Empirical evaluation and comparison of LF-MVCS with eight baseline methods on 11 datasets in terms of clustering accuracy (ACC), normulaized mutual information (NMI) and Purity.

Dataset	A-MKKM	SB-KKM	MKKM [6]	RMKKM [3]	MKKM-MR [11]	ONKC [12]	MVC-LFA [16]	SMKKM [10]	Ours
ACC(%)									
Flower17	51.29	42.41	43.84	49.78	57.66	53.40	59.86	58.40	**67.79**
Flower102	26.85	33.22	22.51	29.19	40.32	38.51	**42.32**	42.26	42.17
PFold	29.05	33.53	27.56	27.81	35.15	36.03	32.47	34.84	**38.34**
Digit	88.82	73.05	47.20	42.85	90.34	90.83	88.83	90.34	**94.00**
CCV	19.60	20.14	18.09	15.77	22.39	22.57	26.03	22.35	**26.67**
Cal-5	35.95	35.95	27.67	31.37	36.11	35.90	37.73	36.03	**40.56**
Cal-10	31.79	33.60	22.17	25.69	32.67	32.38	34.08	32.62	**39.42**
Cal-15	30.55	33.34	19.72	21.96	31.80	30.85	33.10	31.47	**37.91**
Cal-20	29.47	33.87	18.45	21.96	31.47	30.02	32.36	31.24	**37.23**
Cal-25	29.15	33.44	17.26	22.47	30.47	29.33	31.50	30.91	**37.12**
Cal-30	28.53	33.80	16.45	20.46	30.44	29.11	31.39	30.75	**37.03**
NMI(%)									
Flower17	49.87	45.08	44.65	50.41	56.07	52.59	57.90	56.66	**65.74**
Flower102	45.94	48.76	42.69	48.68	56.75	55.19	56.88	58.55	**60.35**
PFold	40.32	41.09	38.27	37.73	44.30	44.92	42.03	44.39	**49.50**
Digit	80.74	66.26	48.74	47.74	83.14	83.70	80.78	83.31	**90.66**
CCV	16.79	17.67	15.08	11.52	18.55	18.57	20.06	18.29	**23.37**
Cal-5	70.26	70.06	66.01	65.94	70.29	70.24	70.99	70.33	**72.65**
Cal-10	61.77	62.78	55.50	55.99	62.23	61.98	63.04	62.22	**65.88**
Cal-15	57.20	58.86	49.35	50.43	57.99	57.39	58.95	57.97	**62.13**
Cal-20	54.23	56.49	45.48	47.43	55.50	54.38	56.07	55.39	**59.53**
Cal-25	52.09	54.50	42.35	45.28	53.10	52.32	54.02	53.46	**57.46**
Cal-30	49.95	53.39	40.08	43.12	51.54	50.45	52.45	51.95	**56.50**
Purity(%)									
Flower17	52.26	44.62	45.30	51.03	59.13	54.24	61.30	59.42	**68.95**
Flower102	32.17	38.46	27.93	34.61	46.36	44.30	48.69	48.58	**50.24**
PFold	37.32	39.19	34.31	33.29	42.44	43.10	39.37	41.62	**45.60**
Digit	88.82	73.40	50.05	45.85	90.34	90.83	88.83	90.34	**94.24**
CCV	23.71	23.34	22.31	20.23	24.70	25.64	28.51	25.47	**29.93**
Cal-5	37.32	37.36	28.42	33.14	37.63	37.27	39.48	37.50	**42.38**
Cal-10	33.62	35.83	23.53	27.16	34.83	34.37	36.26	34.54	**41.73**
Cal-15	32.27	35.34	21.16	24.12	33.71	32.73	35.12	33.27	**40.47**
Cal-20	31.50	35.80	20.07	24.07	33.41	31.93	34.40	33.28	**40.09**
Cal-25	31.12	35.35	18.79	23.92	32.64	31.58	33.93	33.11	**39.85**
Cal-30	30.49	35.71	17.83	22.55	32.42	31.03	33.58	32.75	**39.89**

4.2 Comparison with State-of-the-Art Algorithms

In the experiments, LF-MVCS is compared with the following state-of-the-art multi-view clustering methods.

- **Average multiple kernel k-means (A-MKKM)**: The kernel matrix which is a linear combination of the base kernels with the same coefficients, is taken as the input of standard kernel k-means algorithm.

- **Single best kernel k-means (SB-KKM)**: Standard kernel k-means is performed on each single kernel and the best result is outputted.
- **Multiple kernel k-means (MKKM)** [6]: The algorithm jointly optimizes the consensus clustering indicator matrix and the kernel coefficients.
- **Robust multiple kernel k-means (RMKKM)** [3]: RMKC learns a robust low-rank kernel for clustering by capturing the structure of noises in multiple kernels.
- **Multiple kernel k-means with matrix-induced regularization (MKKM-MR)** [11]: The algorithm introduces a matrix-induced regularization to reduce the redundancy and enhance the diversity of the kernels.
- **Optimal neighborhood kernel clustering with multiple kernels (ONKC)** [12]: ONKC allows the optimal kernel to reside in the neighborhood of linear combination of base kernels and effectively enlarges the searching space of the optimal kernel.
- **Multi-view Clustering via Late Fusion Alignment Maximization (MVC-LFA)** [16]: MVC-LFA proposes to maximally align the consensus clustering indicator matrix with the weighted base clustering indicator matrix.
- **Simple multiple Kernel k-means (SMKKM)** [13]: SimpleMKKM reformulates the problem in MKKM as a minimization-maximization problem in the kernel coefficients and the consensus clustering indicator matrix.

4.3 Experimental Settings

In the experiments, all base kernels are first centered and then normalized so that for all sample $\mathbf{x}_i$ and the p-th kernel, we have $K_p(\mathbf{x}_i, \mathbf{x}_i) = 1$. For all data sets, the true number of clusters is known and we set it to be the true number of classes. In addition, the parameters of RMKKM, MKKM-MR and MVC-LFA are selected by grid search according to the suggestions in their papers. We set the allowable error ϵ to 1×10^{-4} as the termination condition of Algorithm 1.

The widely used clustering accuracy (ACC), normalized mutual information (NMI) and purity are applied to evaluate each algorithm's clustering performance. For all algorithms, we repeat each experiment 50 times with random initialization to reduce randomness caused by k-means in the final spectral clustering, and report the average result. All the experiments are performed on a desktop with Intel(R) Core(TM)-i7-7820X CPU and 64G RAM.

4.4 Experimental Results

Table 2 reports the clustering performance of the above mentioned algorithms on all data sets. From these results, we have the following observations:

- The proposed algorithm consistently demonstrates the best NMI on all data sets. For example, it exceeds the second best one (MVC-LFA) by over 7.8% on Flower17. Also, the superiority of our algorithm is confirmed by the ACC and purity reported in Table 2.

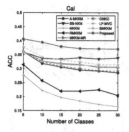 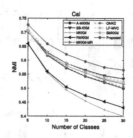 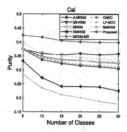

Fig. 2. ACC, NMI and purity comparison with variation of number of classes on Cal. (left) **ACC**, (mid) **NMI** and (right) **purity**

- As a strong baseline, the recently proposed SMKKM [13] outperforms other early-fusion multiple kernel clustering methods in comparison. The proposed algorithm significantly and consistently outperforms SMKKM by **9.08%**, **1.80%**, **5.11%**, **7.35%**, **5.08%**, **2.32%**, **3.66%**, **4.16%**, **4.14%**, **4.00%** and **4.55%** in terms of NMI on all the datasets in the order shown in Table 2.
- The late fusion algorithm MVC-LFA also outperforms most other algorithms, but this algorithm needs to take a lot of effort to determine the best hyperparameters. In some real applications, tuning the parameter may be impossible because there is no ground truth to optimize it. Our LF-MVCS is not only parameter-free, but also outperforms MVC-LFA by **7.84%**, **3.47%**, **7.47%**, **9.88%**, **3.31%**, **1.66%**, **2.84%**, **3.18%**, **3.46%**, **3.44%** and **4.05%** in terms of NMI on all the datasets in the order shown in Table 2.

We also study the clustering performance on Cal of each algorithm concerning the number of classes, as shown in Fig. 2. As observed, our algorithm consistently keeps on the top of all sub-figures when the number of classes varies, indicating the best performance.

In summary, our proposed LF-MVCS demonstrates superior clustering performance over all the SOTA multi-view clustering algorithms and has no hyperparameter to be tuned. From the above experiments, we conclude that the proposed algorithm effectively learns the information from all the base clustering indicator matrices, bringing significant improvements to clustering performance.

4.5 Algorithm Convergence and Runtime

Algorithm Convergence. We have theoretically analyzed the convergence of our algorithm. Figure 4 reflects the variation of the objective function (9) with the number of iterations on four data sets, such as Flower17, ProteinFold, Digit, and Cal-30. It can be seen that the objective function can achieve convergence with very few iterations, and shows that fewer optimization variables can make the algorithm converge faster, making our algorithm more efficient than other algorithms.

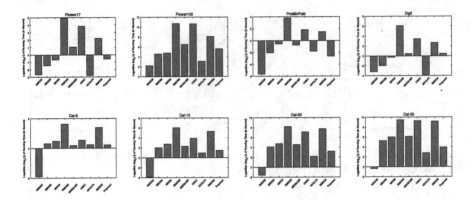

Fig. 3. Runtime of different algorithms on eight benchmark datasets (in seconds). The experiments are conducted on a PC with Intel(R) Core(TM)-i7-7820X CPU and 64G RAM in the MATLAB environment. LF-MVCS is comparably fast to alternatives while providing superior performance and requiring no hyper-parameter tuning. Results for other datasets are omitted due to space limitations.

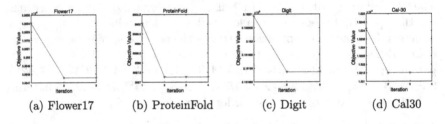

(a) Flower17   (b) ProteinFold   (c) Digit   (d) Cal30

Fig. 4. The objective value of our algorithm on four data sets at each iteration.

Runtime. To compare the computational efficiency of the proposed algorithm, we record the running time of various algorithms on the eight benchmark datasets and report them in Fig. 3. Although the computational complexity of proposed LF-MVCS is $\mathcal{O}(n^2)$, as observed, LF-MVCS does not significantly increase the computational complexity of existing MVC-LFA whose computation complexity is $\mathcal{O}(n)$.

5   Conclusion

This paper proposes a simple but effective algorithm LF-MVCS, which learns a consensus similarity matrix from all the base clustering indicator matrices. We discover that each base clustering indicator matrix is a better feature representation than the original feature and kernel. Based on this, we derive a simple novel optimization goal for multi-view clustering, which learns a consensus similarity matrix for better clustering performance. The proposed algorithm is developed to solve the resultant optimization efficiently. LF-MVCS outperforms all the SOTA methods significantly in the clustering performance. Besides, our algorithm also reduce the computational cost on benchmark datasets in comparison with SOTA methods. We plan to extend our algorithm to a general framework, and use it as a platform to revisit existing late fusion multi-view algorithms and further improve the fusion method.

Acknowledgment. This work was supported by the National Natural Science Foundation of China (project No. 61319020).

References

1. Boyd, S., Vandenberghe, L.: Convex Optimization. Cambridge University Press, Cambridge (2004)
2. Dhillon, I., Guan, Y., Kulis, B.: Kernel k-means: spectral clustering and normalized cuts, pp. 551–556 (2004)
3. Du, L., et al.: Robust multiple kernel k-means using l21-norm (2015)
4. Gao, H., Nie, F., Li, X., Huang, H.: Multi-view subspace clustering. In: Proceedings of the IEEE International Conference on Computer Vision (ICCV), pp. 4238–4246 (2015)
5. Gönen, M., Margolin, A.A.: Localized data fusion for kernel k-means clustering with application to cancer biology. In: Advances in Neural Information Processing Systems (NeurIPS), vol. 27 (2014)
6. Huang, H., Chuang, Y., Chen, C.: Multiple kernel fuzzy clustering. IEEE Trans. Fuzzy Syst. **20**, 120–134 (2012)
7. Kumar, A., Rai, P., Daume, H.: Co-regularized multi-view spectral clustering. Adv. Neural Inf. Process. Syst. (NeurIPS) **24**, 1413–1421 (2011)
8. Li, M., Liu, X., Wang, L., Dou, Y., Yin, J., Zhu, E.: Multiple kernel clustering with local kernel alignment maximization. In: Proceedings of the Twenty-Fifth International Joint Conference on Artificial Intelligence (IJCAI), pp. 1704–1710 (2016)
9. Li, Y., Nie, F., Huang, H., Huang, J.: Large-scale multi-view spectral clustering via bipartite graph. In: Proceedings of the Twenty-Ninth AAAI Conference on Artificial Intelligence (AAAI), pp. 2750–2756 (2015)
10. Liu, X., et al.: Efficient and effective regularized incomplete multi-view clustering. IEEE Trans. Pattern Anal. Mach. Intell. (TPAMI) 3778–3784 (2020)
11. Liu, X., Dou, Y., Yin, J., Wang, L., Zhu, E.: Multiple kernel k-means clustering with matrix-induced regularization. In: Proceedings of the Thirtieth Conference on Artificial Intelligence (AAAI), pp. 1888–1894 (2016)
12. Liu, X., et al.: Optimal neighborhood kernel clustering with multiple kernels. In: Proceedings of the Thirty-First Conference on Artificial Intelligence (AAAI), pp. 2266–2272 (2017)
13. Liu, X., Zhu, E., Liu, J.: SimpleMKKM: simple multiple kernel k-means (2020)
14. Ng, A.Y., Jordan, M.I., Weiss, Y.: On spectral clustering: analysis and an algorithm. Adv. Neural Inf. Process. Syst. (NeurIPS) **13**, 849–856 (2002)
15. Schölkopf, B., Smola, A., Müller, K.R.: Nonlinear component analysis as a kernel eigenvalue problem. Neural Comput. **10**, 1299–1319 (1998)
16. Wang, S., et al.: Multi-view clustering via late fusion alignment maximization. In: Proceedings of the Twenty-Eighth International Joint Conference on Artificial Intelligence (IJCAI), pp. 3778–3784 (2019)

17. Zhang, C., et al.: Generalized latent multi-view subspace clustering. IEEE Trans. Pattern Anal. Mach. Intell. (TPAMI) **42**, 86–99 (2020)
18. Zhang, C., Hu, Q., Fu, H., Zhu, P., Cao, X.: Latent multi-view subspace clustering. In: 2017 IEEE Conference on Computer Vision and Pattern Recognition (CVPR), pp. 4333–4341 (2017)
19. Zhou, S., et al.: Multi-view spectral clustering with optimal neighborhood Laplacian matrix. In: Proceedings of the Thirty-Four AAAI Conference on Artificial Intelligence (AAAI), pp. 6965–6972 (2020)

Minimal Residual Based Iterative Methods and Its Parallel Implementation for Sparse Linear Systems

Jiang Liu[1,2] and Jin Wang[1,2(✉)]

[1] Chongqing Institute of Green and Intelligent Technology,
Chinese Academy of Sciences, Chongqing, China
{liujiang,wangjin}@cigit.ac.cn
[2] University of Chinese Academy of Sciences, Beijing, China

Abstract. Large sparse linear equations often appear in mathematics, physics, computer science, and engineering. Many methods were proposed for various classes of coefficient matrices to solve them with rapid speed and high precision. In this article, we give a new iterative approach called a minimal residual splitting method (MRSM) to solve such systems. It converges for any initial vectors if the coefficient matrix is positive definite but unnecessarily symmetric. Besides, it is easy to implement in a parallel way. And also, we study a generic preconditioner of MRSM. Surprisingly, we get a byproduct formula that solves linear systems through the iterative method by using eigenvalues of the coefficient matrix. In the numerical experiments, we implement these new methods in the GPU platform to study the performances.

Keywords: Iterative methods · Sparse linear system · Minimal residual methods · Parallel algorithms

1 Introduction

In the field of scientific computing, questions are often modeled by Partial Differential Equations (PDEs), which are very difficult to solve, even have no analytical solution. Fortunately, we do not need analytical solutions in many cases, and numerical methods are enough for practical usage [1]. The typical way of numerical methods for PDEs is to discretize them. The linear systems that arise from these discretizations are generally large and sparse, see chapter 2 of [22]. The smaller the discretization error, the larger the dimension of the system, that is, the more significant number of unknowns the equations have. In these linear systems, just a few coefficients are non-zero, that is why we call them large sparse linear systems.

A system of linear equations is usually written as

$$Ax = b, \tag{1}$$

© Springer Nature Singapore Pte Ltd. 2021
K. He et al. (Eds.): NCTCS 2020, CCIS 1352, pp. 87–104, 2021.
https://doi.org/10.1007/978-981-16-1877-2_7

where $A = (a_{ij})$ is an $m_1 \times m_2$ matrix called coefficient matrix, b is an m_1-dimensional vector called right-hand side vector, and x is m_2-dimensional vector called vector of unknowns. In this paper, we just consider the square matrix i.e. $m_1 = m_2 = n$. Furthermore, we assume that A and b are real matrices and vector, respectively. If A is nonsingular matrix, there is a unique solution for the Eq. (1), denoted by $x_* = A^{-1}b$. When the dimension n is very large, computing the inverse matrix A^{-1} is not acceptable for practical usage, as well as other direct methods such as Gaussian elimination, LU factorization, etc.; see [10]. Instead of direct methods, many iterative methods have been proposed [9,14,25,27]. These methods start with an arbitrary vector x_0 and find a sequence of x_k to approximate x_*, where k is $0, 1, 2, \ldots$. Compared with direct methods, the iterative methods have the following advantages, see chapter 11 of [22] and chapter 7 of [2]: (1) they do not need to change the coefficient matrix A in the solving process; (2) they are far easier to implement on parallel computers; (3) it may take less time to solve a system by using a proper iterative method than using any direct methods. A big advantage of iterative methods is that they often can speed up the computing process through a parallel implementation, especially for sparse linear systems [22].

Iterative methods can be divided into two categories: stationary methods and nonstationary methods, see [6]. Stationary iterative methods can be expressed as

$$x_{k+1} = Gx_k + g, \tag{2}$$

where neither G nor g depends on the iteration count k. Widely used stationary methods are the Jacobi method [18], the Richardson method [20], the Gauss-Seidel iterative method [17], the successive overrelaxation method [13,26], etc. Given $x_k = (x_{k,i})$, the Jacobi method updates x_{k+1} by

$$x_{k+1,i} = \frac{1}{a_{ii}} \left(b_i - \sum_{j \neq i} a_{ij} x_{k,j} \right), \tag{3}$$

or rewrite it as matrix form

$$x_{k+1} = D^{-1}(L + U)x_k + D^{-1}b, \tag{4}$$

where $D, -L, -U$ are the diagonal, strict lower part and strict upper part of A, see chapter 4 of [22]. The Richardson method is also a simple iterative method, which updates x_{n+1} by

$$x_{k+1} = x_k + \omega(b - Ax_k). \tag{5}$$

The parameter ω in the above formula is a constant, see [20].

Nonstationary methods differ from stationary methods in that the parameters change at each iteration. Well-known nonstationary methods are projection methods [8]. Denote $\mathcal{K}_k$ and $\mathcal{L}_k$ the search space and the constraint space respectively, and a projection method can be described as

$$\textit{Find } x_k \in x_0 + \mathcal{K}_k, \textit{ such that } b - Ax_k \perp \mathcal{L}_k, \tag{6}$$

in which the approximate solution x_k is found in the affine space $x_0 + \mathcal{K}_k$ and the new residual vector is required orthogonal to the subspace $\mathcal{L}_k$. If each $\mathcal{K}_k$ is Krylov subspace, it's called Krylov subspace methods [15], such as conjugate gradient method [16,24], generalized minimal residual method [21], biconjugate gradient method [12], etc. If $\mathcal{K}_k$ is spaned by just one vector, it's called one-dimensional projection methods, such as the steepest descent method [19] and minimal residual (MR) method [23].

Every iterative method works well in a specific class of linear systems. For instance, the conjugate gradient method is the right choice for symmetric positive definite equations, and the biconjugate gradient method is applicable for general equations. For unsymmetric positive definite equations, an efficient iterative method is called the Hermitian/skew-Hermitian splitting (HSS) iteration method [3]. Furthermore, based on the HSS iteration method, the preconditioned Hermitian/skew-Hermitian splitting (PHSS) iteration method was proposed in [5]. Furthermore, Zhong-Zhi Bai et al. employed different methods to solve the Hermitian part and skew-Hermitian part of equations and get the IHSS (CG, Lanczos) method and the IHSS (CG, CGNE) method [4].

In this paper, we are interested in one-dimensional projection methods. Precisely, based on matrix splitting and minimal residual method, we will develop a so-called minimal residual splitting method (MRSM) for the linear systems with a positive definite coefficient matrix.

2   New Iterative Methods

In this section, we introduce the minimal residual splitting method (MRSM). The main idea is to make residual vectors shortest in various matrix splitting iterative methods. For MRSM, we will analyze the convergence and its parallel implementation. Furthermore, we will study the preconditioning method for it. In particular, we propose the Jacobi preconditioner for MRSM.

2.1   Minimal Residual Splitting Method

For linear system (1), a matrix splitting of its coefficient matrix A is written as

$$A = M - N \tag{7}$$

in which $A \in \mathbb{R}^{n \times n}, M \in \mathbb{R}^{n \times n}$, and A and M are nonsingular. Linear system (1) is equivalent to

$$x = M^{-1}Nx + M^{-1}b. \tag{8}$$

So the below iterative method can be constructed

$$x_{k+1} = M^{-1}Nx_k + M^{-1}b, \tag{9}$$

where x_k is the k-th step approximate solution, and x_0 is an initial vector. The choices of the matrix M give the corresponding stationary iterative methods.

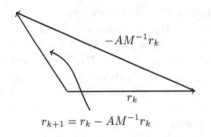

$$r_{k+1} = r_k - AM^{-1}r_k$$

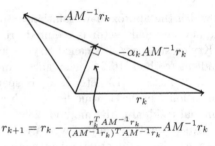

$$r_{k+1} = r_k - \frac{r_k^T AM^{-1}r_k}{(AM^{-1}r_k)^T AM^{-1}r_k} AM^{-1}r_k$$

Fig. 1. Residual vector updating formula in stationary iterative methods

Fig. 2. Shortest residual vector is the perpendicular one

For example, the matrix splittings are $A = D - (L + U)$ in Jacobi method and $A = \frac{1}{\omega}I - (\frac{1}{\omega}I - A)$ in Richardson method.

The k-th step residual vector is

$$r_k = b - Ax_k \tag{10}$$

and the $(k + 1)$-th step residual vector is

$$r_{k+1} = b - Ax_{k+1} = b - A(M^{-1}Nx_k + M^{-1}b), \tag{11}$$

then substitute the N with $M - A$, we obtain

$$r_{k+1} = r_k - AM^{-1}r_k. \tag{12}$$

The figure of the residual vector updating formula (12) is shown as Fig. 1.

In the traditional matrix splitting iterative method such as Jacobi and Richardson method, the residuals (see Fig. 1) are generally not the shortest residual vector (see Fig. 2) which is perpendicular to the vector $-AM^{-1}r_k$. No matter the minimal residual method or steepest descend method, we always make the residual having the least norm along the given direction. In order to get the shortest residual vector in the matrix splitting method, we just introduce a parameter α_k to residual updating formula (12)

$$r_{k+1} = r_k - \alpha_k AM^{-1}r_k, \tag{13}$$

And the approximate solution updating formula accordingly becomes

$$x_{k+1} = x_k + \alpha_k M^{-1}r_k. \tag{14}$$

The next question is how to decide the parameter α_k. First of all, we have an easy observation of the relation between the residues r_k and eigenvectors.

Proposition 1. *If r_k is an eigenvector of AM^{-1}, we just take $\alpha_i = 1/\lambda$, where λ is the eigenvalue associated with r_k, then $r_{k+1} = 0$.*

It is natural to ask how about the relation between the eigenvalues and the solution. Based on the residual formula (14), we can get the following residue.

$$r_{k+1} = \left(\prod_{i=0}^{k} (I_{n \times n} - \alpha_i AM^{-1}) \right) r_0. \tag{15}$$

Let λ_i, for $i = 0, \cdots, n-1$, be the eigenvalues of AM^{-1}. When AM^{-1} is invertible, $\lambda_i \neq 0$ for all i. So, if we choose α_i so that $\{\alpha_i \mid i = 0, \cdots, n-1\} = \{1/\lambda_i \mid i = 0, \cdots, n-1\}$, then we have

Theorem 1. *For arbitrary x_0, it always has*

$$r_n = \left(\prod_{i=0}^{n-1} (I_{n \times n} - \alpha_i AM^{-1}) \right) r_0 = \left(\prod_{i=0}^{n-1} (I_{n \times n} - 1/\lambda_i AM^{-1}) \right) r_0$$

$$= \frac{1}{\prod_{i=0}^{n-1} \lambda_i} \cdot \left(\prod_{i=0}^{n-1} (\lambda_i I_{n \times n} - AM^{-1}) \right) r_0 = \mathbf{0}. \tag{16}$$

In particular, when $M = I$, starting with arbitrary initial point we can get the exact solution by using eigenvalues of coefficient matrix A. Accordingly, the formula is like

$$x_n = x_{n-1} + \alpha_{n-1} r_{n-1} = x_{n-2} + \alpha_{n-2} r_{n-1} + \alpha_{n-1} r_{n-1}$$
$$= x_{n-2} + (\alpha_{n-2} + \alpha_{n-1}) r_{n-1} = x_{n-2} + (\alpha_{n-2} + \alpha_{n-1})(I - \alpha_{n-2}A) r_{n-2}$$
$$= \cdots$$
$$= x_0 + f(A, \alpha_0, \cdots, \alpha_{n-1}, x_0) \tag{17}$$

wherein $1/\alpha_i$ for $0 \leq i \leq n-1$ are eigenvalues of A, $f(A, \alpha_0, \cdots, \alpha_{n-1}, x_0)$ is a polynomial of A, $1/\alpha_i$, and x_0. This is a surprising result. To the best of our knowledge, it is the first time to reveal this relation. However, it might not be applied to solve linear system in practice because computing eigenvalues usually is not easier than solving linear system directly.

Now, we are going to study how to get the minimal residuals. If the iteration count is k, and we assume that x_k and r_k are known, the square of Euclidean norm of r_{k+1} is

$$\|r_{k+1}\|_2^2 = (r_k - \alpha_k AM^{-1} r_k)^T (r_k - \alpha_k AM^{-1} r_k)$$
$$= r_k^T r_k - \alpha_k (AM^{-1} r_k)^T r_k - \alpha_k r_k^T AM^{-1} r_k + \alpha_k^2 (AM^{-1} r_k)^T AM^{-1} r_k$$
$$= r_k^T r_k - 2\alpha_k r_k^T AM^{-1} r_k + \alpha_k^2 (AM^{-1} r_k)^T AM^{-1} r_k. \tag{18}$$

It's a quadratic function about the parameter α_k. To make r_{k+1} shortest, the quadratic function of α_k have to get the minimum, and obviously, the minimum is obtained when

$$\alpha_k = \frac{r_k^T AM^{-1} r_k}{(AM^{-1} r_k)^T AM^{-1} r_k}. \tag{19}$$

Algorithm 1. The Minimal Residual Splitting Method (MRSM)

Input: $A, M, b, x_0, \varepsilon, MaxSteps$

1: $r_0 \leftarrow b - Ax_0$
2: $k \leftarrow 0$
3: **while** $k < MaxSteps$ **and** $\|r_k\|_2 \geq \varepsilon$ **do**
4:     $\alpha_k \leftarrow \dfrac{r_k^T AM^{-1} r_k}{(AM^{-1} r_k)^T AM^{-1} r_k}$
5:     $x_{k+1} \leftarrow x_k + \alpha_k M^{-1} r_k$
6:     $r_{k+1} \leftarrow r_k - \alpha_k AM^{-1} r_k$
7:     $k \leftarrow k + 1$
8: **end while**
9: **return** x_k

The figure of the residual updating formula (13) with $\alpha_k = \dfrac{r_k^T AM^{-1} r_k}{(AM^{-1} r_k)^T AM^{-1} r_k}$ is shown as Fig. 2.

Therefore, the algorithm can be represented as Algorithm 1, called the minimal residual splitting method (MRSM). The MRSM is a one-dimensional projection method. In k-step iteration, the search direct is $M^{-1} r_k$, so the search space $\mathcal{K}_k = span\{M^{-1} r_k\}$. And the shortest r_{k+1} is get when $r_{k+1} \perp AM^{-1} r_k$, so the constraint space $\mathcal{L}_k = span\{AM^{-1} r_k\}$.

In Algorithm 1, there are 2 matrix-vector multiplications, 3 inner products, and 2 vector additions in a loop. Therefore, if it converges in N iterations, the computational costs are $2N$ matrix-vector multiplications, $3N$ inner products, and $2N$ vector additions.

Let $A = L + D + R$ be a splitting of A as before. When $M = D$, $M = D - L$, and $M = I$, correspondingly, $\alpha_k = 1$, $\alpha_k = \omega$, the result method respectively are Jacobi, Richardson and Gauss-Seidel methods. Locally speaking, the MRSM approximates to a solution faster than these methods, which is one point why we are interested in this new iterative method. However, the global convergence is not simple, even divergent. We will analyze the convergence after we introduce a new type of precondition method for MRSM.

2.2   Preconditioning

Preconditioning is used to reduce the condition number of the coefficient matrix and improve the efficiency and robustness of iterative methods. In this paper, preconditioning is also used to make iterative methods fitted for positive definite equations. Preconditioners can be applied from the left and the right. Besides, split preconditioning is also possible. In this section, we discuss the left preconditioned methods of the Algorithm 1.

If $P \in \mathbb{R}^{n \times n}$ is a nonsingular matrix, the Eq. (1) is equivalent to

$$PAx = Pb. \tag{20}$$

They have the same solution and (20) may be easier to solve. Select a matrix splitting of the matrix PA,

Algorithm 2. The Left Preconditioned Minimal Residual Splitting Method (PMRSM)

Input: $A, M, P, b, x_0, \varepsilon, MaxSteps$
1: $r_0 \leftarrow Pb - PAx_0$
2: $k \leftarrow 0$
3: **while** $k < MaxSteps$ **and** $\|r_k\|_2 \geq \varepsilon$ **do**
4: $\quad \alpha_k \leftarrow \dfrac{r_k^T PAM^{-1}r_k}{(PAM^{-1}r_k)^T PAM^{-1}r_k}$
5: $\quad x_{k+1} \leftarrow x_k + \alpha_k M^{-1}r_k$
6: $\quad r_{k+1} \leftarrow r_k - \alpha_k PAM^{-1}r_k$
7: $\quad k \leftarrow k + 1$
8: **end while**
9: **return** x_k

$$PA = M - N. \tag{21}$$

The residual vector of Eq. (20) is

$$r_k = Pb - PAx_k. \tag{22}$$

Apply the minimal residual splitting method to the left preconditioned Eq. (20), which is substituting the A in the Algorithm 1 by PA and substituting the b in the Algorithm 1 by Pb, and we obtain an algorithm described as the Algorithm 2.

Naturally, the Algorithm 2 is also a one-dimensional projection method whose search space $\mathcal{K}_k = span\{M^{-1}r_k\}$ and constraint space $\mathcal{L}_k = span\{PAM^{-1}r_k\}$. Different matrices P and M will give different methods, for example, if we let $P = M = I$ then we get the primitive MR method [23].

Compared with the Algorithm 1, there are 1 more matrix-vector multiplication per iteration in Algorithm 2. But it will generally converge in less iterations than the Algorithm 1. If we suppose it converges in N iterations, the computational costs are $3N$ matrix-vector multiplications, $3N$ inner products, and $2N$ vector additions.

2.3 Convergence Analysis

In the k-th step of the Algorithm 2, if the r_k is not zero and perpendicular to $PAM^{-1}r_k$, α_k will be zero, and r_{k+1} will be equals to r_k. In such situation, the PMRSM can't improve the precision of the approximate solution no matter how many iterations are done. To guarantee PMRSM working, we must make sure that for all $k = 0, 1, 2, \ldots$, there is no k making r_k perpendicular to $PAM^{-1}r_k$. Fortunately, if PAM^{-1} is positive definite but not necessarily symmetric, the angles between r_k and $AM^{-1}r_k$ are always sharp angles or zero, and the method will be convergent.

Theorem 2. *If PAM^{-1} is real positive definite, the Algorithm 2 is convergent for any $\varepsilon > 0$ under ignoring "MaxSteps".*

Proof. There are two cases. (I) If there exists a k, $r_k = 0$, the method is convergent after k step iteration. (II) If for all k, $r_k \neq 0$, we have $r_k^T PAM^{-1} r_k > 0$, and so

$$\|r_{k+1}\|_2^2 = r_k^T r_k - \frac{(r_k^T PAM^{-1} r_k)^2}{(PAM^{-1} r_k)^T PAM^{-1} r_k} < \|r_k\|_2^2. \tag{23}$$

It means that the sequence $\{\|r_k\|_2^2\}$ is strictly descending, and since $\|r_k\|_2^2 \geq 0$, the sequence must have a limit when $k \to \infty$. The critical point is whether the limit is zero. In fact, the zero limit can be proved by contradiction. We assume that the limit $\lim_{k \to \infty} r_k$ is not zero vector. By (23), $\|r_k\|_2 < \|r_0\|_2$ for all r_k, that is, all r_k are in the compact set $\{X \in \mathbb{R}^n \mid \|X\|_2 \leq \|r_0\|_2\}$. Therefore, both sets $\{r_k^T PAM^{-1} r_k \mid k = 0, 1, 2, \dots\}$ and $\{r_k^T (PAM^{-1})^T (PAM^{-1}) r_k \mid k = 0, 1, 2, \dots\}$ are compact, and then $\lim_{k \to \infty} r_k^T PAM^{-1} r_k$ must have a supremum and $\lim_{k \to \infty} r_k^T (PAM^{-1})^T (PAM^{-1}) r_k$ must have an infimum. In consequence, there are some positive real numbers c and d such that

$$\lim_{k \to \infty} r_k^T PAM^{-1} r_k = c > 0, \tag{24}$$

and

$$\overline{\lim_{k \to \infty}} \, r_k^T (PAM^{-1})^T (PAM^{-1}) r_k = d > 0. \tag{25}$$

Therefore, there exists a positive integer N_1, such that for any $k > N_1$,

$$|r_k^T PAM^{-1} r_k - c| < \frac{c}{2}, \tag{26}$$

so

$$r_k^T PAM^{-1} r_k > \frac{c}{2}. \tag{27}$$

And there exists a positive integer N_2, such that for any $k > N_2$,

$$|r_k^T (PAM^{-1})^T (PAM^{-1}) r_k - d| < \frac{d}{2}, \tag{28}$$

and then

$$r_k^T (PAM^{-1})^T (PAM^{-1}) r_k < \frac{3d}{2}. \tag{29}$$

Let's take the integer $N := \max\{N_1, N_2\}$, then for any $k > N$,

$$\frac{r_k^T PAM^{-1} r_k}{r_k^T (PAM^{-1})^T (PAM^{-1}) r_k} > \frac{c}{2} \Big/ \frac{3d}{2} = \frac{c}{3d} > 0. \tag{30}$$

In consequence,

$$\begin{aligned} \|r_{k+1}\|_2^2 &= r_k^T r_k - \frac{(r_k^T PAM^{-1} r_k)^2}{(PAM^{-1} r_k)^T PAM^{-1} r_k} \\ &< r_k^T r_k - \frac{c}{3d} < \cdots < r_N^T r_N - \frac{c}{3d}(k - N), \end{aligned} \tag{31}$$

therefore,

$$\lim_{k \to \infty} \|r_k\|_2^2 = -\infty. \tag{32}$$

It's in contradiction with the truth that $\|r_k\|_2^2 \geq 0$. So $\lim_{k\to\infty} \|r_k\|_2^2 = 0$, and the method is convergent. □

From the proof, it can be easily seen that the matrix PAM^{-1} being positive definite is sufficient to the convergence. When a linear system and a splitting of the corresponding coefficient matrix are given, the speed of convergence can be inferred from the following theorem [22].

Theorem 3. *If PAM^{-1} is real positive definite, the residual vectors generated by the Algorithm 2 satisfy the relation*

$$\|r_{k+1}\|_2 \leq \sqrt{1 - \frac{\lambda_{min}^2(PAM^{-1} + (PAM^{-1})^T)}{4\|PAM^{-1}\|_2^2}} \|r_k\|_2, \tag{33}$$

where $\lambda_{min}(X)$ is the minimal eigenvalue of the matrix X and $\|PAM^{-1}\|_2$ is the norm of PAM^{-1}.

Proof. Expand the square of the Euclidean norm of the residual vector $r_{k+1} = r_k - \alpha_k PAM^{-1} r_k$ as

$$\begin{aligned}
\|r_{k+1}\|_2^2 &= (r_{k+1}, r_{k+1}) = (r_{k+1}, r_k - \alpha_k PAM^{-1} r_k) \\
&= (r_{k+1}, r_k) - \alpha_k(r_{k+1}, PAM^{-1} r_k).
\end{aligned} \tag{34}$$

Naturally, the shortest r_{k+1} is get when r_{k+1} is perpendicular to $PAM^{-1} r_k$, so $(r_{k+1}, PAM^{-1} r_k) = 0$. Therefore,

$$\begin{aligned}
\|r_{k+1}\|_2^2 &= (r_{k+1}, r_k) = (r_k - \alpha_k PAM^{-1} r_k, r_k) \\
&= (r_k, r_k) - \alpha_k(PAM^{-1} r_k, r_k) \\
&= \|r_k\|_2^2 \left[1 - \frac{(PAM^{-1} r_k, r_k)}{(PAM^{-1} r_k, PAM^{-1} r_k)} \cdot \frac{(PAM^{-1} r_k, r_k)}{(r_k, r_k)} \right] \\
&= \|r_k\|_2^2 \left[1 - \frac{(PAM^{-1} r_k, r_k)^2}{(r_k, r_k)^2} \cdot \frac{\|r_k\|_2^2}{\|PAM^{-1} r_k\|_2^2} \right].
\end{aligned} \tag{35}$$

By using the Courant-Fischer min-max theorem, see 4.2 of [17],

$$\begin{aligned}
\frac{(PAM^{-1} r_k, r_k)}{(r_k, r_k)} &= \frac{(\frac{PAM^{-1} + (PAM^{-1})^T}{2} r_k, r_k)}{(r_k, r_k)} \\
&\geq \lambda_{min}\left(\frac{PAM^{-1} + (PAM^{-1})^T}{2} \right).
\end{aligned} \tag{36}$$

And

$$\|PAM^{-1} r_k\|_2 \leq \|PAM^{-1}\|_2 \|r_k\|_2. \tag{37}$$

So

$$\|r_{k+1}\|_2^2 \le \left(1 - \frac{\lambda_{min}^2(PAM^{-1} + (PAM^{-1})^T)}{4\|PAM^{-1}\|_2^2}\right)\|r_k\|_2^2. \tag{38}$$

Let's consider the square root of the above inequality. If $1 - \dfrac{\lambda_{min}^2(PAM^{-1} + (PAM^{-1})^T)}{4\|PAM^{-1}\|_2^2}$ is always non-negative, then inequality (33) is obtained. By using the Cauchy–Schwarz inequality [11], we can show the positivity as follows. First,

$$(PAM^{-1}r_k, r_k)^2 \le (PAM^{-1}r_k, PAM^{-1}r_k)(r_k, r_k). \tag{39}$$

And then

$$0 \le \frac{(PAM^{-1}r_k, r_k)}{(PAM^{-1}r_k, PAM^{-1}r_k)} \cdot \frac{(PAM^{-1}r_k, r_k)}{(r_k, r_k)} \le 1. \tag{40}$$

Therefore,

$$0 \le \frac{\lambda_{min}^2(PAM^{-1} + (PAM^{-1})^T)}{4\|PAM^{-1}\|_2^2} \le 1. \tag{41}$$

So $1 - \dfrac{\lambda_{min}^2(PAM^{-1} + (PAM^{-1})^T)}{4\|PAM^{-1}\|_2^2}$ must be non-negative.   □

Let $c = \sqrt{1 - \dfrac{\lambda_{min}^2(PAM^{-1} + (PAM^{-1})^T)}{4\|PAM^{-1}\|_2^2}}$. By using the Theorem 3, we obtain $\|r_k\|_2 \le c^k\|r_0\|_2$. So the Algorithm 2 is convergent if $c < 1$. If PAM^{-1} is positive definite, the matrix $PAM^{-1} + (PAM^{-1})^T$ will be symmetric positive definite. So the minimal eigenvalue of $PAM^{-1} + (PAM^{-1})^T$ is greater than zero. Therefore, $c < 1$ and the Algorithm 2 is convergent. In addition, the less c is, the faster the algorithm converge. This provides us some instructions how to select the matrix P.

Notice that the Theorem 3 gives an upper bound of the Euclidean norm of residual vectors. In practice, the iteration may converge much faster than this upper bound.

2.4   Preconditioned Minimal Residual Methods

The Theorem 2 tells us the Algorithm 2 is convergent if PAM^{-1} is positive definite. So we need to select P and M carefully in order to guarantee the convergence. When A is a symmetric positive matrix, there is a unitary matrix U such that $U^T AU$ is a diagonal matrix. In such case, if we take $P = U^T$ and $M^{-1} = U$, then the convergent rate is controlled by the condition number of $U^T AU$. Inspired by such choice, we may consider nonsingular matrices $P = M^{-1}$ or $P^T = M^{-1}$. When $P^T = M^{-1}$, the matrix PAM^{-1} is positive definite if and only if A is positive definite, because if $r_k \ne 0$, $Pr_k \ne 0$, and $(Pr_k)^T A(Pr_k) > 0$. Therefore, this suggests us choose P and M such that $P^T = M^{-1}$, then the convergence of PMRSM will be determined by the definite positivity of A.

Algorithm 3. The Jacobi-Preconditioned Minimal Residual Method (JMRM)

Input: $A, b, x_0, \varepsilon, MaxSteps$

1: $r_0 \leftarrow \sqrt{D^{-1}}b - \sqrt{D^{-1}}Ax_0$
2: $k \leftarrow 0$
3: **while** $k < MaxSteps$ **and** $\|r_k\|_2 \geq \varepsilon$ **do**
4:     $\alpha_k \leftarrow \dfrac{r_k^T \sqrt{D^{-1}}A\sqrt{D^{-1}}r_k}{(\sqrt{D^{-1}}A\sqrt{D^{-1}}r_k)^T \sqrt{D^{-1}}A\sqrt{D^{-1}}r_k}$
5:     $x_{k+1} \leftarrow x_k + \alpha_k \sqrt{D^{-1}}r_k$
6:     $r_{k+1} \leftarrow r_k - \alpha_k \sqrt{D^{-1}}A\sqrt{D^{-1}}r_k$
7:     $k \leftarrow k + 1$
8: **end while**
9: **return** x_k

Based on the choice $P^T = M^{-1}$, we could fix P first. Then M can be computed correspondingly. Look at Algorithm 2, in the computing process, M^{-1} is actually required rather than M. Therefore, P can be picked freely, except it must be invertible. Let's recall various preconditioning techniques for linear systems, such as Jacobi preconditioner, incomplete LU factorization, and sparse approximate inverse; see [7]. Jacobi preconditioner is the simplest preconditioner among them. It is just the diagonal part of A, which is usually denoted D. But neither AD^{-1} nor $D^{-1}A$ is positively definite. To settle down the positivity, we let $P = M^{-1} = \sqrt{D^{-1}}$, then $\sqrt{D^{-1}}A\sqrt{D^{-1}}$ is positive definite, because the diagonal elements of a positively definite A must be positive numbers, so $\sqrt{D^{-1}}$ always exists and is nonsingular. Thus, we can obtain a method described as the Algorithm 3 called Jacobi-Preconditioned Minimal Residual Method (JMRM), which is convergent when A is positive definite.

In Algorithm 3, we can save $\sqrt{D^{-1}}$ in a vector, and the cost of the matrix-vector multiplication decreases to the same cost as vector addition. In general, vector addition is faster than matrix-vector multiplication. So there are just 1 matrix-vector multiplication, 3 inner products, and 4 vector additions per loop. And the computational costs are N matrix-vector multiplications, $3N$ inner products, and $4N$ vector additions, if the Algorithm 3 converges in N loops.

Certainly, there may have a better choice of preconditioner than Jacobi's preconditioner for a typical system. For instance, it may be better to choose P or P^T as the lower or upper triangular matrix of A in some cases. Anyway, we should consider the corresponding condition numbers, which affect the convergent rate. Furthermore, we should be aware of the time consuming when P is a complicated matrix. In the future, we will further study the selection of P as $P^T = M^{-1}$, even P and M simultaneously for $P^T \neq M^{-1}$.

2.5 Parallel Implementation Analysis of MRSMs

The MRSM can be easily carried out in a parallel way, which is another important reason why we are interested in it. In Algorithm 1, there are just some basic operations of matrices and vectors including matrix-vector multiplications, vector inner products, and vector additions. The main idea is to implement these operations in parallel way.

In Algorithm 1, for a while loop, it needs n many inner products of n-dimensional vectors to compute $M^{-1}r_k$ and so for computing $AM^{-1}r_k$; thus it needs $2n + 2$ many such inner products for computing α_k; the previous computation $M^{-1}r_k$ can be used for x_{k+1}; similarly the previous computation $AM^{-1}r_k$ can be used for r_{k+1}. Totally, it needs $2n + 2$ many inner products in each while loop. From Algorithm 1, we can see these inner products can be computed in a parallel way. That is, if $M^{-1} := \begin{pmatrix} M_1 \\ \vdots \\ M_n \end{pmatrix}$ where $M_1, \cdots, M_n$ are row vectors of M^{-1}, then $M^{-1}r_k$ can be computed by parallelly computing $M_i r_k$ for $1 \leq i \leq n$. It is similar for computing $AM^{-1}r_k$.

The parallel implementation can speed up computing, especially for sparse matrix. For dense matrices A and M^{-1}, the total multiplications of numbers is $2n^2 + 2n$ in each wile loop of Algorithm 1. Let's assume the matrices A and M^{-1} are sparse so that each row has at most N many nonzero elements, where N is far smaller than n. For such sparse matrices, in each wile loop of Algorithm 1, the total multiplications of numbers is $2Nn + 2n$, which is far smaller than $2n^2 + 2n$.

3   Numerical Experiment

In this section, we report the results of the numerical experiment. The experiment consists of two parts: (i) the performances of MRSM with $M = I$, $M = D$ and $M = D - L$, and (ii) the performances of the Jacobi method, the minimal residual method, and the Algorithm 3. We implemented these methods with CUDA, whose version is 6.5 under double-precision, and the major calculations are deployed in the GPU. We selected the vector whose elements are all 1.0 as the initial vector $x_0 = [1, 1, \ldots, 1]^T \in \mathbb{R}^n$, and the results of the iteration are stated as follows.

The first one is a sparse linear system that was randomly generated by a program. The coefficient matrix is a 100000×100000 real positive definite matrix. For this matrix, there are about 15 non-zero elements every row, they locate randomly, and their values are generated by a uniform distribution whose range is $[1, 10]$ except the diagonal elements. The maximal and minimal singular values of the coefficient matrix are 206.476 and 4.85812, so its condition number is 42.5012. The right-hand side vector b is a random 100000×1 vector whose elements are generated by a uniform distribution whose range is $[0, 1]$. Besides, there are several linear systems from practice. The information of the coefficient matrices is listed in the Table 1. All the coefficient matrices of these linear systems are unsymmetric positive definite that can not be solved directly by the conjugate gradient method.

Table 1. The information of the coefficient matrices

Name	Size	Condition number	Website
add20	2395	12047.1	https://sparse.tamu.edu/Hamm/add20
add32	4960	136.68	https://sparse.tamu.edu/Hamm/add32
memplus	17758	129435.5	https://sparse.tamu.edu/Hamm/memplus
pde2961	2961	642.5	https://sparse.tamu.edu/Bai/pde2961

3.1   The Minimal Residual Splitting Method

The Jacobi method, the Richardson method, and the Gauss-Seidel method are three usual matrix splitting based methods. We employ the matrix splittings of the three methods to build the MRSM. We test the performances of MRSM, respectively, with $M = I$, $M = D$, and $M = D - L$.

First, the random linear system can not be solved by MRSM with $M = D - L$, because the inverse matrix of $D - L$ has so many non-zero elements that there is not enough memory. Precisely, $(D - L)^{-1}$ has 553742471 non-zero elements and requires 6.19 GB memory under sparse storing, or 18.63 GB memory under full storing. However, k20m has 5 GB memory without a page replacement technique. So we get an error of "out of memory". The Euclidean norm of the residual vectors of MRSM with $M = I$ and $M = D$ are shown as Fig. 3. And the Euclidean norm of the residual vectors of MRSM with $M = I$, $M = D$ and $M = D - L$ are shown as Fig. 4, 5, 6 and Fig. 7. The time elapsed is shown as Table 2.

The MRSM with $M = D$ is simpler than the MRSM at $M = D - L$, but the performance is higher in many linear systems. The MRSM with $M = D - L$ has a fatal shortcoming that $(D - L)^{-1}$ is too hard to compute, and the number of non-zero elements of $(D - L)^{-1}$ is usually too large. As a result, the time consuming for computing $(D - L)^{-1}$ may be longer than for solving the linear system by other methods, and there may not be enough memory to save $(D - L)^{-1}$ like the random linear system. In some systems, the norm of the residual vectors of MRSM with $M = D$ and $M = D - L$ does not descend after some iterations. Because AM^{-1} is an indefinite matrix, and the minimal and maximal eigenvalues of $AM^{-1} + (AM^{-1})^T$ of some coefficient matrices are listed in the Table 3. This is the most important reason why the MRSM may not be used to solve positive definite linear systems.

3.2   The Jacobi-Preconditioned Minimal Residual Method

The Algorithm 3 can be regarded as the combination of the Jacobi method and minimal residual (MR) method, so it is meaningful to test its improvement. We will give the performance of the Jacobi method, the minimal residual method, and the Algorithm 3. The norms of the residual vectors in the iterations are shown as Figs. 8, 9, 10, 11 and 12, and the time elapsed is shown as Table 4. In these figures, the curves of the Jacobi method are bolder than others, because

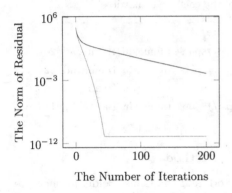

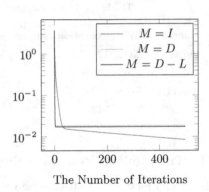

Fig. 3. The norm of residual vectors of the random equation solved by MRSM

Fig. 4. The norm of residual vectors of equation add20 solved by MRSM

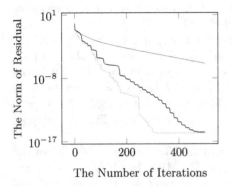

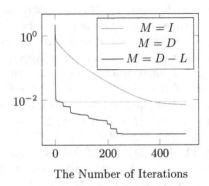

Fig. 5. The norm of residual vectors of equation add32 solved by MRSM

Fig. 6. The norm of residual vectors of equation memplus solved by MRSM

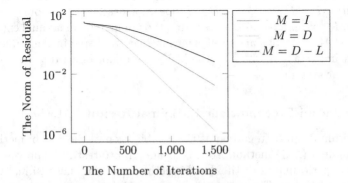

Fig. 7. The norm of residual vectors of equation pde2961 solved by MRSM

Table 2. The time elapsed of MRSM

Equation	Iterations count	$M = I$(s)	$M = D$(s)	$M = D - L$(s)
random	200	0.151917	0.157593	–
add20	500	0.0518402	0.0532239	0.0714827
add32	500	0.0529995	0.0555897	0.0596154
memplus	500	0.080645	0.0844801	0.205035
pde2961	1500	0.149676	0.159099	0.619093

Table 3. The minimal and maximal eigenvalues of $AM^{-1} + (AM^{-1})^T$

Equation	Matrix B	$\lambda_{max}(B^T + B)$	$\lambda_{min}(B^T + B)$
add20	AD^{-1}	4.3684	−0.368403
add20	$A(D - L)^{-1}$	6.23755	−0.248754
add32	AD^{-1}	4.16651	−0.166987
memplus	AD^{-1}	7.99835	−3.99835
memplus	$A(D - L)^{-1}$	9.26045	−3.26487

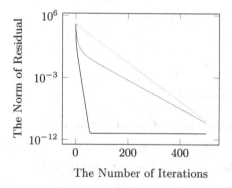

Fig. 8. The norm of residual vectors of the random equation solved by MR, the Jacobi method (JM), and JMRM

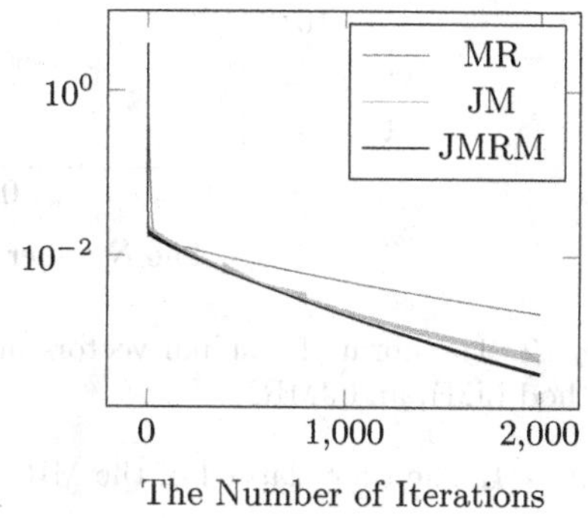

Fig. 9. The norm of residual vectors of equation add20 solved by MR, the Jacobi method (JM), and JMRM

the Jacobi method is not stable and oscillates in the local areas. Nevertheless, from the Table 4, we find the Algorithm 3 always spends more time than the other two methods, so it is necessary to reduce the number of iterations to speed up.

From the results, the solutions are given by Algorithm 3 is obviously more accurate after the same number of iterations, and more stable than the Jacobi method. However, Algorithm 3 has a disadvantage that the amount of computation in every iteration is more considerable than the other two methods.

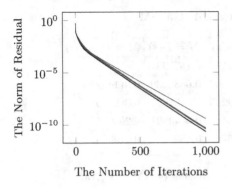

Fig. 10. The norm of residual vectors of equation add32 solved by MR, the Jacobi method (JM), and JMRM

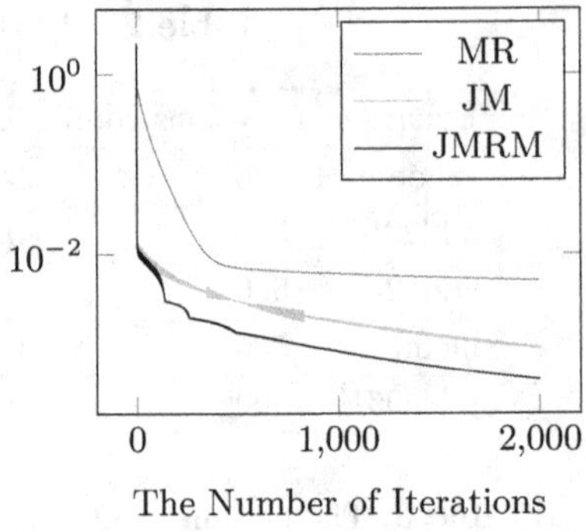

Fig. 11. The norm of residual vectors of equation memplus solved by MR, the Jacobi method (JM), and JMRM

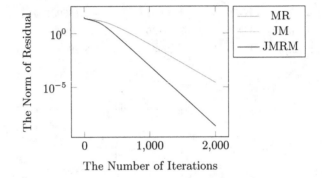

Fig. 12. The norm of residual vectors of equation pde2961 solved by MR, the Jacobi method (JM), and JMRM

Table 4. The time elapsed of the MR method, the Jacobi method, and the JMRM

Equation	Iterations count	MR(s)	the Jacobi method(s)	Algorithm 3(s)
random	500	0.380	0.356	0.413
add20	2000	0.110314	0.0603942	0.121341
add32	1000	0.110314	0.0603942	0.121341
memplus	2000	0.328938	0.227821	0.387009
pde2961	2000	0.197177	0.10962	0.229265

4   Conclusion

Minimal residual iterative methods can be developed in different ways that can numerically solve positive definite linear systems, particularly on unsymmetric positive definite equations that cannot be solved by the conjugate gradient method. However, the Algorithm 3 behaves just in low condition number equations and is not high-performance enough when the coefficient matrix is ill-conditioned, so we must develop other preconditioners to improve performance. The major problem is that the traditional preconditioning techniques may not be suitable for this new methodology, because we need further consider the matrix splitting of PA, which is not considered in the traditional preconditioners for unsymmetric linear systems. In future work, we may develop more preconditioners based on existing preconditioning techniques.

Acknowledgement. The authors also would like to thank Dr. Chengling Fang for the discussion on proof of convergence theorem and noticing the residual formula (16). This work was partially supported by NSF of China (No. 61672488), the equipment union project of Academy of Chinese Sciences (No. 6141A01140104), National Key R&D Program of China (No. 2018YFC0116704).

References

1. Adsuara, J., Cordero-Carrión, I., Cerdá-Durán, P., Aloy, M.: Scheduled relaxation Jacobi method: improvements and applications. J. Comput. Phys. **321**, 369–413 (2016). https://doi.org/10.1016/j.jcp.2016.05.053
2. Ascher, U.M., Greif, C.: A first course on numerical methods. Computational Science and Engineering, vol. 7. SIAM (2011)
3. Bai, Z.Z., Golub, G.H., Ng, M.K.: Hermitian and skew-Hermitian splitting methods for non-Hermitian positive definite linear systems. J. Matrix Anal. Appl. **24**(3), 603–626 (2003). https://doi.org/10.1137/S0895479801395458
4. Bai, Z.Z., Golub, G.H., Ng, M.K.: On inexact Hermitian and skew-Hermitian splitting methods for non-Hermitian positive definite linear systems. Linear Algebr. Appl. **428**(2–3), 413–440 (2008). https://doi.org/10.1016/j.laa.2007.02.018
5. Bai, Z.Z., Golub, G.H., Pan, J.Y.: Preconditioned Hermitian and skew-Hermitian splitting methods for non-Hermitian positive semidefinite linear systems. Numerische Mathematik **98**(1), 1–32 (2004). https://doi.org/10.1007/s00211-004-0521-1
6. Barrett, R., et al.: Templates for the solution of linear systems: building blocks for iterative methods. Society for Industrial and Applied Mathematics (1994). https://doi.org/10.1137/1.9781611971538, https://epubs.siam.org/doi/abs/10.1137/1.9781611971538
7. Benzi, M.: Preconditioning techniques for large linear systems: a survey. J. Comput. Phys. **182**(2), 418–477 (2002). https://doi.org/10.1006/jcph.2002.7176
8. Brezinski, C.: Projection methods for linear systems. J. Comput. Appl. Math. **77**(1–2), 35–51 (1997). https://doi.org/10.1016/S0377-0427(96)00121-5
9. Chae, M., Walker, S.G.: An EM-based iterative method for solving large sparse linear systems. Linear Multilinear Algebr. **68**(1), 45–62 (2020). https://doi.org/10.1080/03081087.2018.1498061
10. Davis, T.A., Rajamanickam, S., Sid-Lakhdar, W.M.: A survey of direct methods for sparse linear systems. Acta Numerica **25**, 383–566 (2016). https://doi.org/10.1017/S0962492916000076
11. Dragomir, S.: A survey on Cauchy-Bunyakovsky-Schwarz type discrete inequality. JIPAM J. Inequal. Pure Appl. Math. **4**(3), 1–142 (2003)

12. Fletcher, R.: Conjugate gradient methods for indefinite systems. In: Watson, G.A. (ed.) Numerical Analysis. LNM, vol. 506, pp. 73–89. Springer, Heidelberg (1976). https://doi.org/10.1007/BFb0080116

13. Frankel, S.P.: Convergence rates of iterative treatments of partial differential equations. Math. Tables Other Aids Comput. 4(30), 65–75 (1950). https://doi.org/10.2307/2002770, www.jstor.org/stable/2002770

14. Freund, R.W., Nachtigal, N.M.: QMR: a quasi-minimal residual method for non-Hermitian linear systems. Numerische Mathematik 60(1), 315–339 (1991). https://doi.org/10.1007/BF01385726

15. Gutknecht, M.H.: A brief introduction to Krylov space methods for solving linear systems. In: Kaneda, Y., Kawamura, H., Sasai, M. (eds.) Frontiers of Computational Science. Springer, Heidelberg (2007). https://doi.org/10.1007/978-3-540-46375-7_5

16. Hestenes, M.R., Stiefel, E.: Methods of conjugate gradients for solving linear systems, vol. 49. NBS, Washington, DC (1952). https://doi.org/10.6028/jres.049.044

17. Horn, R.A., Johnson, C.R.: Matrix Analysis, 2nd edn. Cambridge University Press, USA (2012)

18. Jacobi, C.G.: Ueber eine neue auflösungsart der bei der methode der kleinsten quadrate vorkommenden lineären gleichungen. Astronomische Nachrichten 22(20), 297–306 (1845). https://doi.org/10.1002/asna.18450222002

19. Osadcha, O., Marszalek, Z.: Comparison of steepest descent method and conjugate gradient method (2017). http://ceur-ws.org/Vol-1853/p01.pdf

20. Richardson, L.F.: The approximate arithmetical solution by finite differences of physical problems involving differential equations, with an application to the stresses in a masonry dam. Philos. Trans. R. Soc. Lond. Ser. A 210(459–470), 307–357 (1911). https://doi.org/10.1098/rsta.1911.0009

21. Saad, Y., Schultz, M.H.: GMRES: a generalized minimal residual algorithm for solving nonsymmetric linear systems. SIAM J. Sci. Stat. Comput. 7(3), 856–869 (1986). https://doi.org/10.1137/0907058

22. Saad, Y.: Iterative methods for sparse linear systems, vol. 82. SIAM (2003). https://doi.org/10.1137/1.9780898718003

23. Sheng, X., Su, Y., Chen, G.: A modification of minimal residual iterative method to solve linear systems. Math. Probl. Eng. 2009, 9 (2009). https://doi.org/10.1155/2009/794589

24. Shewchuk, J.R.: An introduction to the conjugate gradient method without the agonizing pain. Report, Carnegie Mellon University (1994)

25. Sleijpen, G.L., Fokkema, D.R.: BiCGstab(l) for linear equations involving unsymmetric matrices with complex spectrum. Electron. Trans. Numer. Anal. 1(11), 11–32 (1993)

26. Varga, R.S.: Matrix Iterative Analysis. Springer, Heidelberg (1962). https://doi.org/10.1007/978-3-642-05156-2

27. Van der Vorst, H.A.: Bi-CGSTAB: a fast and smoothly converging variant of Bi-CG for the solution of nonsymmetric linear systems. SIAM J. Sci. Stat. Comput. 13(2), 631–644 (1992). https://doi.org/10.1137/0913035

Deep Learning

A Simple yet Effective Unsupervised Adversarial Example Generation Framework for Vulnerability Assessment on Deep Learning

Rongkun Zhang, Dan Liu$^{(\boxtimes)}$, Wentao Zhao, Qiang Liu, and Chengzhang Zhu

College of Computer, National University of Defense Technology, Changsha, China
`danliu@nudt.edu.cn`

Abstract. Generating adversarial examples to discover the vulnerability of deep learning model is critical yet challenging. Although adversarial sample generation methods have attracted the attention of lots of recent research, they heavily depend on supervised label information. As a result, they could only work on a specific classification task and fail to evaluate the vulnerability of unsupervised deep learning models. In this paper, we propose a simple yet effective unsupervised adversarial example generation framework for vulnerability assessment of deep learning. This framework achieves an effective unsupervised adversarial example generation by jointly maximizing the distribution divergence of the deep learning model-generated representations of regular examples and adversarial examples and minimizing the dissimilarity between regular examples and adversarial examples in the original space. In this way, the adversarial examples generated by this framework can mislead downstream tasks while camouflaging themselves. We further implement this framework to an adversarial image generation methods to demonstrate the performance of the proposed framework. Extensive experiments on four benchmarks demonstrate that the proposed method can find the vulnerability of the LeNet5 (a widely used deep learning model) much more effective even compared to state-of-the-art supervised competitors.

Keywords: Adversarial examples · Unsupervised learning · Deep learning vulnerability assessment

1 Introduction

Deep Learning (DL) is powerful yet vulnerable for a variety of machine learning tasks. Although DL achieves the state-of-the-art machine learning performance, it would report unexpected results when affecting by even small perturbations [23]. Recently, many efforts [4,8,11,12,17,24,26] have been paid on generating adversarial samples (the samples that can mislead classifiers to make a wrong classification) to exploit the vulnerability of DL methods, which could further contribute to improve DL's robustness [9].

© Springer Nature Singapore Pte Ltd. 2021
K. He et al. (Eds.): NCTCS 2020, CCIS 1352, pp. 107–122, 2021.
https://doi.org/10.1007/978-981-16-1877-2_8

Original Images   Perturbed Images by FGSM   Perturbed Images by PGD   Perturbed Images by Our Method

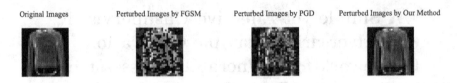

Fig. 1. A demonstration of adversarial examples w.r.t. different generating methods.

Current adversarial example generation methods follow two paradigms. The first paradigm draws adversarial samples from a learned distribution that can mislead specific classifiers [1, 27]. The methods in this paradigm (distribution-based methods) can generate an infinite number of adversarial examples. However, these examples may have large diverse with regular examples. This diversity is easy to recognized and has been well studied by robust learning methods. As a result, the methods in the first paradigm currently have limited contribution to finding new vulnerabilities of a deep learning model. The second paradigm generates adversarial samples by slightly disturbing regular examples [3, 7, 9, 15, 19, 23]. Typically, the methods in this paradigm (perturbation-based methods) learn a small perturbation that can confuse a classifier when the perturbation is added to regular examples. The adversarial examples generated by these methods are hard to be defended because they are very similar to regular examples. Thus, they have an excellent opportunity to discover new vulnerabilities of a deep learning model.

Although the perturbation-based methods have attracted increasing attention in recent research, all of them are supervised adversarial example generation methods. In other words, these methods require annotated labels to train how to generate adversarial samples. For example, if a perturbation-based method needs to mislead a cat-or-dog classifier, this method needs to know whether the generated perturbation on a cat can let the classifier predict a dog. The supervised adversarial example generation methods are sufficient for specific classification tasks; however, the adversarial examples generated by these methods are hard to be migrated to other tasks. However, evaluating deep learning vulnerability expects this migration ability because most powerful deep learning models are pre-trained in some large data sets and then fine-tuned for a specific task with a small number of supervised examples. Empowered by the migration ability, adversarial examples may find in-depth vulnerabilities of deep learning models caused by different training stages. Also, the supervised adversarial example generation methods cannot work when annotation labels are missing. This missing widely exists in learning tasks, such as unsupervised classification, clustering, similar example retrieval. These tasks also require robust deep learning model, but none of the existing adversarial example generation method can be used to validate the vulnerability of a deep learning model for them.

Furthermore, human beings can identify the adversarial examples generated by the existing perturbation-based methods, especially when the content of the example is single and pure (e.g., face recognition, product classification, and fingerprint identification). For example, we can easily recognize the middle two pictures in

Fig. 1 (generated by the state-of-the-art methods [9,19]) are disturbed examples. As a result, these adversarial examples are also not effected for vulnerability evaluation. Human beings can easily recognize the adversarial examples generated by these methods because they regulate only the sum of perturbation. Accordingly, some parts of an example may be very different from the regular one.

In this paper, we propose a simple yet powerful unsupervised adversarial example generation framework (UAEG for short) to tackle the above problems. The intuition of the proposed framework is to maximize the distribution divergence of the DL-generated representations of the regular examples and adversarial examples while minimizing the dissimilarity of regular examples and their corresponding adversarial examples in their original space. By maximizing the distribution divergence in the representation space, adversarial examples will show different behaviors in the downstream tasks; thus, the adversarial examples generated by the proposed framework can mislead various classification tasks on the same deep learning networks. Furthermore, maximizing the distribution divergence does not dependence on label information. Consequently, the proposed framework can generate adversarial examples in an unsupervised fashion. By minimizing the dissimilarity between regular examples and their corresponding adversarial examples, the generated adversarial examples would camouflage themselves to evade robust deep learning models or defense methods. In this way, the proposed framework can generate more effective adversarial examples with a higher probability of discovering the vulnerability of deep learning models.

The UAEG framework can be implemented for various types of data with different distribution divergence measures and data similarity measures. In this paper, we implement the proposed framework to an adversarial image generation method to demonstrate its performance. Specifically, we introduce mutual information neural estimation (MINE) [2] as the distribution divergence measurement. The MINE method could provide an approximate KL-divergence of two distributions, which is differentiable for easy optimization. We adopt a L_2 norm to derive the dissimilarity measure of images in their original space. The L_2-norm-derived dissimilarity has a good property that can highlight the significant difference in an image matrix. As a result, minimizing the L_2-norm-derived dissimilarity of an adversarial example and a regular example will force the adversarial example similar to the regular example in each pixel, which is hard for human beings to recognize.

The contribution of this work can be summarized as follows:

- We propose a simple yet powerful adversarial example generation framework. As far as we know [18,22], UAEG is the first unsupervised framework for adversarial example generation, which proposes a general approach for vulnerability evaluation of deep learning models in various tasks.
- We implement the UAEG framework to an unsupervised adversarial image generation method, namely UAEG-Image. The UAEG-Image method adopts a novel distribution divergence measurement that seamlessly integrates into the proposed framework to provide unsupervised learning ability. It also introduces a L_2-norm-derived dissimilarity measure catering to the data characteristics of images, which camouflages adversarial examples to evade detection.

Extensive experiments on well-known benchmarks show the superior performance of the proposed method to generate effective adversarial examples for vulnerability evaluation of deep learning models. Specifically, the adversarial examples generated by the proposed method reduces the classification accuracy of a famous deep learning model, LeNet-5 [16], from 95.37% to 57.13% on KMNIST [5], from 62.22% to 34.75% on CIFAR10 [13,14] (grayscale), and from 89.07% to 15.97% on FashionMNIST [25]. The above performance reduction is even significantly large than the two state-of-the-art supervised adversarial example generation methods [9,19].

The rest of this paper is organized as follows: Sect. 2 reviews the related work. Section 3 introduces the proposed framework and its implementation. Section 4 presents the experiments and analyzes the experimental results. Finally Sect. 5 concludes this paper and discusses some promising directions that can be worked in future.

2   Related Work

2.1   Adversarial Examples

After the phenomenon of weakness in the DL-based image classifier model [23] was found, more and more similar situations arose quickly in other fields [4,12, 17,24], although similar threats have existed mentioned in the work [18]. We take image classification tasks as examples like the work [23] to show adversarial examples. Given an image sample x, the purpose is to find another image sample x', which is similar to x and satisfy Eq. (1). The x' is an adversarial example, and η presents the perturbation. The $\|\cdot\|_2$ presents the Euclidean norm, which can measure the distance between x and x'. Minimizing the distance means maximizing the similarity between x and x'. The metric of distance(or similarity) is not limited to $\|\cdot\|_2$. It can be other methods depending on the circumstance of a study. Let $C(\cdot)$ notes the classifier, which outputs the label of input. In our work, we still pay attention to neural networks based image classification models.

$$minimize \quad \|\eta\|_2$$
$$s.t. \quad \eta = x - x' , \ C(x) \neq C(x') , \ x' \in [0,1]^n \tag{1}$$

2.2   Adversaries Capabilities

According to the degrees of the knowledge of DL models which adversaries can get, adversaries capabilities can be divided into three kinds of situation. In the white-box case, the adversaries can get all knowledge of a DL model, such as specific structure, hyper-parameters, weights/bias of the model, also the adversaries have the full right to operate the training data. In the black-box case, the adversaries can get no knowledge at all. They can only use their inputs and the outputs corresponding to the DL model. In the semi-black-box case, which is between in white-box and black-box, the adversaries can get partial knowledge of a DL model or only can get the training dataset. Considering we leverage our framework on image classification tasks and need to capture the feature of the images through a pre-trained neural network classifier, hence our experiment is under a white-box scenario.

2.3 Methods of Generating Adversarial Examples

Fast Gradient Sign Method. Also called FGSM [9], given an input image x, this method utilizes Eq. (2) to generate adversarial example x', where y is the label associated with x, $J(x, y)$ is the loss used to train the classifier, $\nabla_x J(x, y)$ is the gradient of $J(x, y)$ w.r.t. x, ϵ is a hyper-parameter to adjust the size of perturbation. This method is a kind of one-step method.

$$x' = x + \epsilon \cdot sign(\nabla_x J(x, y)) \tag{2}$$

Projected Gradient Descent. Compared with FGSM [9], PGD [19] perturbs input image x many steps with small step size α. and it is a more potent variant FGSM [9]. This method utilizes Eq. (3) to fabricate x'. The Γ_ϵ is the operation of projection, which limits the x' in a ϵ size of the distance of x.

$$x'_{t+1} = \Gamma_\epsilon(x_t + \alpha \cdot sign(\nabla_{x_t} J(x_t, y))) \tag{3}$$

Other Methods. There are also other types of methods to fabricate adversarial examples, such as the Distribution-based methods [1,27]. These method draw adversarial samples from a learned distribution that can mislead specific classifiers.

2.4 Mutual Information and MINE Method

Mutual information (MI) is a measure in Information Theory, which can be used to reveal the uncertainty or dependence between random variables. The Eq. (4) shows the mutual information which can be regarded as the information contained in T about A. Where $p(t, a)$ is the joint distribution of variables, $p(t)$ and $p(a)$ are the edge distribution of variables.

$$I(T; A) := \sum_{t \in T} \sum_{a \in A} p(t, a) log \frac{p(t, a)}{p(t)p(a)} \tag{4}$$

$$I(T; A) = \mathbb{E}_{\mathbb{P}_{TA}}[G_\delta] - log\left(\mathbb{E}_{\mathbb{P}_T \otimes \mathbb{P}_A}\left[e^{G_\delta}\right]\right) \tag{5}$$

In one case [2], mutual information can be used to measure the similarity between distributions, and the researchers proposed a method, namely Mutual Information Neural Estimator (MINE), to estimate the KL-divergence of two distributions which can be approximated to the MI value. The MINE [2] trains a neural network by a function as Eq. (5), which can approximate the MI value. Where $\mathbb{P}_{TA}$ stands for joint distribution and $\mathbb{P}_T \otimes \mathbb{P}_A$ for the product of the marginals, G_δ is a mapping from $\mathcal{T} \times \mathcal{A}$ to $\mathbb{R}$ with the parameters δ. One of our work is to calculate the similarity between distributions in the representation space. Therefore we use the MINE [2] method to measure the distribution divergence.

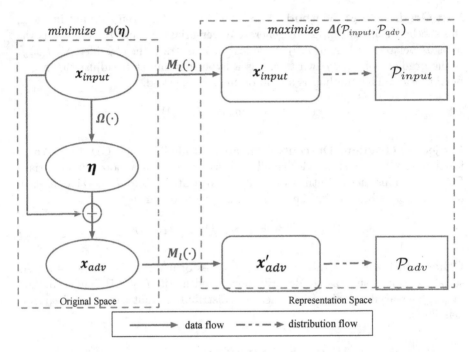

Fig. 2. The Architecture of the UEAG Framework. In this figure, x_{input} and x_{adv} refer to the original input and adversarial example respectively. $\Omega(\cdot)$ stands for the neural network to generate the perturbation η. $M_l(\cdot)$ presents the operation of getting the feature representation x'_{input} and x'_{adv}, which follow the distribution of $\mathcal{P}_{input}$ and $\mathcal{P}_{adv}$ respectively. $\Phi(\cdot)$ presents a mertic in original space, and the $\Delta(\cdot, \cdot)$ presents the distribution divergence measure.

3   Methodology

3.1   The UAEG Framework

The UAEG framework considers the similarity of input data in original space and the similarity of distributions of representations in representation space between original input data and adversarial examples. Hence, it can be abstracted into two parts to generate adversarial examples as Eq. (6), the first(left part) is about the original space part, the second(right part) is about the representation space part, and the α and β are coefficients to balance the weight. As shown in Fig. 2, UAEG works as follows.

$$minimize \quad \alpha\Phi(\boldsymbol{\eta}) - \beta\Delta(\mathcal{P}_{input}, \mathcal{P}_{adv}) \tag{6}$$

In the first part, UAEG focuses on increasing the similarity between original input data $\boldsymbol{x}_{input} \in D_{input}$ and adversarial examples $\boldsymbol{x}_{adv} \in D_{adv}$ in original space. Let $\boldsymbol{\eta}$ be the perturbation, the $\boldsymbol{x}_{input}$ add the $\boldsymbol{\eta}$ becomes the $\boldsymbol{x}_{adv}$. Hence, minimizing the dissimilarity between $\boldsymbol{x}_{input}$ and $\boldsymbol{x}_{adv}$ can be transformed into minimizing the perturbation $\boldsymbol{\eta}$. For the perturbation $\boldsymbol{\eta}$, UAEG uses $\Omega(\boldsymbol{x}_{input})$ to generate, the $\Omega(\cdot)$ can be a neural network model. The $\Phi(\cdot)$ presents a mertic to measure the value of $\boldsymbol{\eta}$. It can be L_0, L_2, L_∞ norm, or others depending on the specific situation.

In the second part, UAEG focuses on decreasing the similarity between the distributions $\mathcal{P}_{input}$ and $\mathcal{P}_{adv}$ in representation space, which means maximizing the distribution divergence. Hence, we use a minus sign in Eq. (6), and the $\Delta(\cdot)$ presents the divergence. Specifically, note that $\boldsymbol{x}'_{input} \in D'_{input} \sim \mathcal{P}_{input}$ and $\boldsymbol{x}'_{adv} \in D'_{adv} \sim \mathcal{P}_{adv}$. First, UAEG leverages $M_l(\boldsymbol{x}_{input})$ and $M_l(\boldsymbol{x}_{adv})$ to get $\boldsymbol{x}'_{input}$ and $\boldsymbol{x}'_{adv}$ respectively, where $M_l(\cdot)$ presents the extraction operation of feature representations of the Model M(e.g., the l-th layer output of a target DL-based model) corresponding to the input. While in practice, it is difficult to identify the distributions specifically in representation space and find substitutions for them to compute the MI value or the divergence. Hence, UAEG introduces the MINE [2] method to design and train a neural network which could provide an approximate divergence of two distributions ($\mathcal{P}_{input}$ and $\mathcal{P}_{adv}$). Therefore UAEG can reduce the similarity between $\mathcal{P}_{input}$ and $\mathcal{P}_{adv}$ by maximizing the divergence through the MINE [2] method without knowing the specific formalization of distributions. Maximizing the divergence means minimizing the MI value between the distributions. Hence, Eq. (6) can also be written as Eq. (7), where the outputs of the $\Psi(\cdot)$ presents the MI value.

$$minimize \quad \alpha\Phi(\boldsymbol{\eta}) + \beta\Psi(\mathcal{P}_{input}, \mathcal{P}_{adv}) \tag{7}$$

Through iterations, UAEG can fabricate adversarial examples under the two parts of constrains above. Besides its novel perspective, UAEG has good expansibility since it can be instantiated to a specific method by specifying its four components Ω, Φ, M, and Ψ. We introduce an instance of UAEG next on image classification tasks and formalization it namely UAEG-Image.

3.2   An UAEG Instance: UAEG-Image

UAEG-Image instantiates UAEG by a neural network which can be used to image transformation task as the function Ω, a L_2 norm metric function Φ, a pre-trained image classifier model M based on the LeNet-5 [16] architecture. The output of function M_l is the l-th layer output of the Model M and a designed neural network S with the objective function Ψ of the MINE [2] method. After specifying these components, UAEG-Image can be formalized according to the UAEG framework.

Formalize the UAEG-Image Method. In the first part, according to Eq. (7), UAEG-Image focuses on increasing the similarity between original input images x_{input} and adversarial examples x_{adv}. UAEG-Image applies L_2 norm as the metric. Specifically, note that the η stands for the discrepancy between the x_{input} and x_{adv}, we transform the L_2 norm operation from the target of x_{input} and x_{adv} to the target of η. Hence, we leverage the L_2-norm-derived metric as the function Φ to minimize the value of $\|\eta\|_2$. For the perturbations η in images, UAEG-Image needs an operation that can make an input image to another image in a different style(e.g., an image as perturbation) and keep the same size. This kind of operation is similar to the image transformation tasks which transform an input image into an output image in a different style [10, 28]. Hence, UAEG-Image utilizes the architecture of the network model in [28] as the function $\Omega(\cdot)$ to generate η. The formalization of this part shows in Eq. (8).

$$\Phi(\eta) = \|\eta\|_2 \tag{8}$$

In the second part, according to Eq. (7) and the MINE [2] method, it can be formalized as Eq. (9). S_θ denotes a designed neural network S parametrized by mapping $S_\theta : D'_{input} \times D'_{adv} \rightarrow R$, and θ represents the weights and bias of S. We can get samples $(x'_{input}, \tilde{x}'_{adv})$ of edge distribution $\mathcal{P}_{input} \otimes \mathcal{P}_{adv}$ by shuffling the samples (x'_{input}, x'_{adv}) from the joint distribution $\mathcal{P}_{input}\mathcal{P}_{adv}$ along the batch axis to estimate the expectations [2].

$$\Psi(\mathcal{P}_{input}, \mathcal{P}_{adv}) = \mathbb{E}_{\mathcal{P}_{input\ adv}}[S_\theta] - log\left(\mathbb{E}_{\mathcal{P}_{input} \otimes \mathcal{P}_{adv}}\left[e^{S_\theta}\right]\right) \tag{9}$$

Launch the UAEG-Image Method. Algorithm 1 shows the primary process of the UAEG-Image Method. Specifically, we train the designed MINE [2] network S through the objective function $\Psi(\mathcal{P}_{input}, \mathcal{P}_{adv})$ to get the MI value according to Eq. (9). Then we fix the parameters θ of S, through minimizing Eq. (7) to training the network $\Omega(\cdot)$. We use the α and β as the hyper-parameters to balance the weight. The t stands for a threshold of $\Phi(\eta)$ or $\Psi(\mathcal{P}_{input}, \mathcal{P}_{adv})$ as a signal to adjust α and/or β. After repeating the steps above, we can get a trained neural network $\Omega(\cdot)$(also can be seen as a function), which can be used to fabricate adversarial examples from input images.

4   Experiments and Evaluation

We utilize the instance UAEG-Image with several experiments on image classification tasks to demonstrate the UAEG framework's effectiveness and performance. First, we show that the UAEG-Image method can fabricate effective adversarial examples to reduce the accuracy of predictions of target classifier on three datasets. Then we compare the UAEG-Image method with FGSM [9] and PGD [19] in different parameters, which shows the UAEG-Image method is better than other methods considering both the performance and visual feelings.

Algorithm 1. The UAEG-Image Method

Input: $X_{input} = (x'^{(1)}_{input}, ..., x'^{(N)}_{input})$(separated to N/m batches, each batch denoted as X_{input}), α and β(the coefficients of the objective function), t(the threshold), γ(the iteration times for the network S training), n(the iteration times for the whole training process).

Output: $\Omega(\cdot)$

1 **for** *iteration = 1 to n* **do**
2   **repeat**
3   /* Begain to train Ψ network */
4   **for** *iterS = 1 to γ* **do**
5              $\eta = \Omega(X_{input})$
6              $X_{adv} = \eta + X_{input}$
7   clip X_{adv} to (0,1)
8              $X'_{input}, X'_{adv} = M_l(X_{input}, X_{adv})$
9   update $S_\theta \leftarrow$

                $\frac{1}{m} \sum_{i=1}^m S_\theta(X'^{(i)}_{input}, X'^{(i)}_{adv}) - log(\frac{1}{m} \sum_{i=1}^m exp(S_\theta(X'^{(i)}_{input}, \widetilde{X}'^{(i)}_{adv})))$
10   **end**
11   /* Training Ψ network end */
12   /* Begain to train Ω network */
13          $\eta = \Omega(x_{input})$
14          $X_{adv} = \eta + X_{input}$
15   clip X_{adv} to (0,1)
16          $X'_{input}, X'_{adv} = M_l(X_{input}, X_{adv})$
17   update $\Omega \leftarrow minimize$ $\alpha\Phi(\eta) + \beta\Psi(\mathcal{P}_{input}, \mathcal{P}_{adv})$
18   /* Training Ω network end */
19   adjust $\alpha, \beta \leftarrow$ compare the threshold t with $\Phi(\eta)$ or $\Psi(\mathcal{P}_{input}, \mathcal{P}_{adv})$
20   **until** *all the batches are finished in one iteration;*
21 **end**

4.1 Data Sets

There are three datasets used to train the classifiers respectively, the KMNIST dataset [5], the CIFAR-10 dataset [13,14] and the FashionMNIST dataset [25]. We use the training set as the training data, then we get three pre-trained classifiers(each classifier can be a target model M) correspondingly. Note that the distribution of the images dataset with three channels is more complicated than that with one channel in representation space. Hence, we transfer the CIFAR-10 dataset [13,14] from RGB to grayscale.

4.2 Parameter Settings

Function M. We utilize the LeNet-5 [16] neural network as the base of pre-trained classifiers. The architecture (which shows in Table 1) consists of two convolutional layers and three fully connected layers. We use all the training data in each dataset to train the classifiers. For the KMNIST dataset [5], we get 95.37% accuracy on the testing dataset. For the CIFAR10 dataset [13,14]

Table 1. Architecture of pre-trained classifiers

Type of layers	KMNIST&FashionMNIST	CIFAR10 to Grayscale
Convolution(Relu)	5×5×6 (padding:same)	5×5×6
MaxPooling	2×2	2×2
Convolution(Relu)	5×5×16	5×5×16
MaxPooling	2×2	2×2
FullyConnected(Relu)	120	120
FullyConnected(Relu)	84	84
FullyConnected	10	10

Table 2. Architecture of the MINE network

Type of layers	For three datasets
FullyConnected(For one input)	5
FullyConnected(For another input)	5
Add(Relu)	5
FullyConnected	1

(we transfer them into grayscale), we get 62.22% accuracy on the testing dataset. For the FashionMNIST dataset [25], we get 89.07% accuracy on the testing dataset.

Function Ω. In the work of [28], the architecture of making new images contains two convolutional layers, 6 or 9 residual blocks and two transposed convolutions. We use the same structure with six residual blocks to generate perturbations η from input images.

Function Ψ. According to the MINE [2] method and the mapping of neural network S_θ in Sect. 3.2, the architecture receives two inputs and output a real number, which shows in Table 2. Therefore, in our work, after getting the representations of input images and adversarial examples in representation space, we send each input to a fully connected layer, then add the outputs of the layer together, finally send the added result to a fully connected structure to get a real number.

4.3 Effectiveness of Unsupervised Adversarial Example Generation

Experimental Settings. We randomly select $N = 10000$ samples from the training dataset for 300 epochs, each epoch contains 10 batches, each batch contains 1000 samples. We take a strategy to train the model function Ω and Ψ. According to Eq. (7), we use $\mathcal{L}_S$ presents the $\Psi(\mathcal{P}_{input}, \mathcal{P}_{adv})$, $\mathcal{L}_{norm2}$ presents the $\Phi(\eta)$, and $\mathcal{L}_G$ presents the $\alpha\Phi(\eta)$ adding $\beta\Psi(\mathcal{P}_{input}, \mathcal{P}_{adv})$. We first fix the coefficient α of the $\mathcal{L}_S$ and set the coefficient β of $\mathcal{L}_{norm2}$ a small value until the

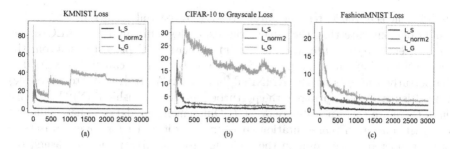

Fig. 3. Illustrations of the training loss of the UAEG-Image method over KMNIST, CIFAR-10(grayscale), and FashionMNIST datasets.

$\mathcal{L}_S$ tends to be zero or below a threshold t, then change α and/or β to a fixed value gradually until the training process finished. In more detail, for KMNIST dataset [5], we set α equals 30, β equals 0.01, and the threshold t is 0.25 as the initial values. Once $\mathcal{L}_S$ below 0.25, we begin to increase β 0.1 per batch from 0.01 to 1.51, once the $\mathcal{L}_{norm2}$ below 6.8, we begin to increase β 0.1 per batch from 1.51 to 3.51, once the $\mathcal{L}_{norm2}$ below 5.2, we begin to increase β 0.1 per batch from 3.51 to 7.01, then fix the coefficients until the training phase finished. For CIFAR-10 dataset [13,14], we transfer the original images to grayscale images as our input images. We set α equals 10, β equals 0.01, and the threshold t is 0.25 as the initial values. Once $\mathcal{L}_S$ below 0.25, we begin to increase β 0.1 per batch from 0.01 to 1.51, then once the $\mathcal{L}_{norm2}$ below 6.8 and the iteration is more than 15, we begin to increase β 0.1 per batch from 1.51 to 3.51, once the $\mathcal{L}_{norm2}$ below 5.2 and the iteration is more than 20, we begin to increase β 0.1 per batch from 3.51 to 7.01, then fix the coefficients until the training phase finished. For FashionMNIST dataset [25], we set α equals 10, β equals 0.01, and the threshold t is 0.25 as the initial values. Once $\mathcal{L}_S$ below 0.25, we begin to increase β 0.1 per batch from 0.01 to 1.51, once the $\mathcal{L}_{norm2}$ below 4.9, we begin to decrease α 0.1 per batch from 10.0 to 5.0, then fix the coefficients until the training phase finished. The change of the loss function can be seen in Fig. 3. The loss curves of $\mathcal{L}_S$ and $\mathcal{L}_{norm2}$ show a downward trend. Because of the adjust of coefficients, the $\mathcal{L}_G$ curve does not always decrease, but after the adjustment, it also shows a downward trend. The trend of the curves meets the optimization expectation of our experiment.

Finding: UAEG-Image Generates Effective Adversarial Images in Unsupervised Fashion. For the KMNIST dataset [5], UAEG-Image can make the accuracy of the testing phase to 57.13% (pre-trained classifier is 95.37%) on the 10000 testing data. We display partial results in Fig. 4 (a). For the CIFAR-10 dataset [13,14], UAEG-Image can make the accuracy of the testing phase to 34.75% (pre-trained classifier is 62.22%) on the 10000 testing data. We display partial results in Fig. 4 (b) For the FashionMNIST dataset [25], UAEG-Image can make the accuracy of the testing phase to 15.87% (pre-trained classifier is

89.07%) on the 10000 testing data. We display partial results in Fig. 5. The results demonstrate that the adversarial examples fabricated by UAEG-Image can efficiently reduce the accuracy of image classifiers' predictions. From the perspective of human visual perception, the perturbations can still be sensed in the perturbed images on the KMNIST [5] and CIFAR10 [13,14] (to grayscale) while it is barely perceptible in the perturbed images on FashionMNIST [25]. We think there are three possible reasons. (i)Different image datasets have different data distribution in representation space, making it easier for some datasets to find a proper balance point in the training process. (ii)For some datasets, the L_2 norm may not be the optimal image similarity measure for human visual perception. (iii)In the training process, the optimal balance point may not be found, and the current result may be trapped in a local solution, which may be related to the adjustment scheme of coefficients like α and β.

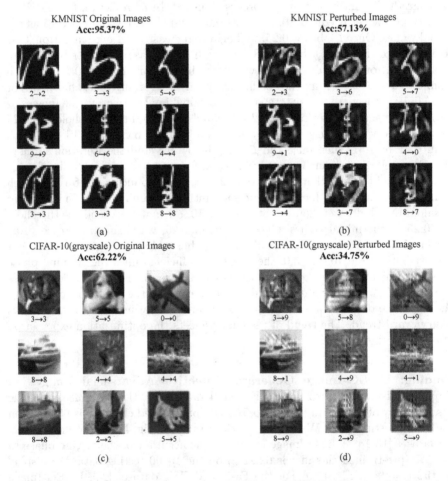

Fig. 4. The results of UAEG-Image over KMNIST and CIFAR-10(grayscale) datasets.

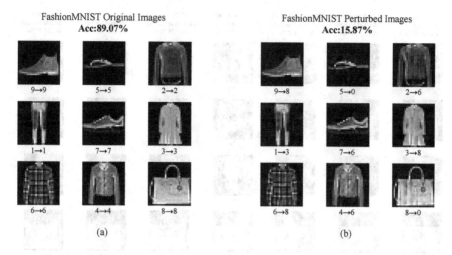

Fig. 5. The results of UAEG-Image over the FashionMNIST dataset.

4.4 Effectiveness of Deep Learning Vulnerability Assessment

Experimental Settings. We compare the UAEG-Image method with two state-of-the-art methods FGSM [9] and PGD [19], which also generate adversarial examples to fool the pre-trained classifier. The pre-trained classifier is the model function M in Sect. 4.2 on FashionMNIST dataset [25]. It is convenient to use open source tools to recurrence the two methods to fabricate adversarial examples, such as CleverHans [20], Foolbox [21], and AdverTorch [6]. We select AdverTorch [6] framework with default parameters doing the comparative experiments. Notably, to compare clearly, we adjust the ϵ of PGD method [19], which clips the perturbations into $(-\epsilon, \epsilon)$ to limit the size of distortion.

Finding: UAEG-Image Enables Better Performance on Deep Learning Vulnerability Assessment Even Compared to Supervised Adversarial Example Generation Methods. We do the comparative experiments among UAEG-Image, FGSM [9], and PGD(with different ϵ) [19] methods on Fashion-MNIST [25]. The comparative results are shown in Fig. 6. We mark the correct class to the identified class under each image(e.g. 9→9). The (a) column images are the original images without perturbations, and the (g) column images are the images perturbed by UAEG-Image. From the (c) column images, although the PGD method [19] with default parameters performs better in accuracy value, the perturbations are very obvious than UAEG-Image. Hence we adjust the ϵ to get the (d) column images, a comparative accuracy result with UAEG-Image, while the perturbations are still evident. When we adjust the ϵ to reduce the perturbations visually, which are the (e) and (f) column images, we find that the accuracy increased than UAEG-Image. For the (b) column images, the FGSM [9] can not generate perturbation for each image, even if add the perturbation, it still apparent visually, and the accuracy is higher than UAEG-Image. Combine

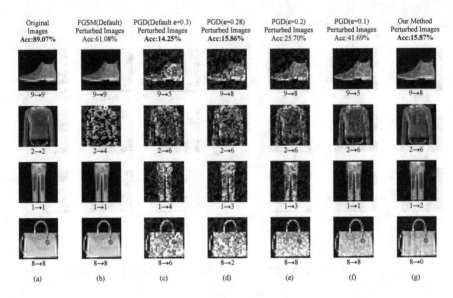

Fig. 6. Comparative results of different methods over the FashionMNIST dataset.

the two aspects(perturbations on the visual and the accuracy), UAEG-Image is better than FGSM [9] and PGD method [19].

5 Conclusion and Future Work

This paper introduces a new framework UAEG to generate adversarial examples and its instance UAEG-Image. They leverage original input data in original space and the distributions of datasets in representation space to efficiently fabricate adversarial examples to reduce the accuracy of the classifiers. Although UAEG-Image works on relatively simple datasets, it obtains better performance compared to its competitors. We are further studying more complex image datasets such as three channels RGB images, or search for another metric way to measure the similarity in original space.

Acknowledgement. The work is supported by National Natural Science Foundation of China under Grant No. U1811462.

References

1. Baluja, S., Fischer, I.: Adversarial transformation networks: learning to generate adversarial examples. arXiv preprint arXiv:1703.09387 (2017)
2. Belghazi, M.I., et al.: Mutual information neural estimation. In: Dy, J.G., Krause, A. (eds.) Proceedings of the 35th International Conference on Machine Learning, ICML 2018, Stockholmsmässan, Stockholm, Sweden, July 10–15, 2018. Proceedings of Machine Learning Research, Vol. 80, pp. 530–539. PMLR (2018)

3. Carlini, N., Wagner, D.A.: Towards evaluating the robustness of neural networks. In: IEEE Symposium on Security and Privacy, pp. 39–57. IEEE (2017). https://doi.org/10.1109/SP.2017.49
4. Chernikova, A., Oprea, A., Nita-Rotaru, C., Kim, B.: Are self-driving cars secure? evasion attacks against deep neural networks for steering angle prediction. In: IEEE Security and Privacy Workshops, SP Workshops 2019, San Francisco, CA, USA, pp. 132–137. IEEE (2019). https://doi.org/10.1109/SPW.2019.00033
5. Clanuwat, T., Bober-Irizar, M., Kitamoto, A., Lamb, A., Yamamoto, K., Ha, D.: Deep learning for classical Japanese literature. arXiv preprint arXiv:1812.01718 (2018)
6. Ding, G.W., Wang, L., Jin, X.: AdverTorch v0.1: an adversarial robustness toolbox based on pytorch. arXiv preprint arXiv:1902.07623 (2019)
7. Dong, Y., et al.: Boosting adversarial attacks with momentum. In: CVPR, pp. 9185–9193. IEEE (2018)
8. Dong, Y., Su, H., Wu, B., Li, Z., Liu, W., Zhang, T., Zhu, J.: Efficient decision-based black-box adversarial attacks on face recognition. In: IEEE Conference on Computer Vision and Pattern Recognition, CVPR 2019, Long Beach, CA, USA, pp. 7714–7722. Computer Vision Foundation/IEEE (2019). https://doi.org/10.1109/CVPR.2019.00790
9. Goodfellow, I.J., Shlens, J., Szegedy, C.: Explaining and harnessing adversarial examples. In: ICLR (Poster) (2015)
10. Isola, P., Zhu, J.Y., Zhou, T., Efros, A.A.: Image-to-image translation with conditional adversarial networks. In: CVPR, pp. 1125–1134 (2017)
11. Jan, S.T.K., Messou, J., Lin, Y., Huang, J., Wang, G.: Connecting the digital and physical world: Improving the robustness of adversarial attacks. In: AAAI, pp. 962–969 (2019)
12. Kakizaki, K., Yoshida, K.: Adversarial image translation: unrestricted adversarial examples in face recognition systems. In: SafeAI@AAAI. CEUR Workshop Proceedings, Vol. 2560, pp. 6–13. CEUR-WS.org (2020)
13. Krizhevsky, A.: Learning multiple layers of features from tiny images (2009)
14. Krizhevsky, A., Nair, V., Hinton, G.: The cifar-10 dataset, http://www.cs.toronto.edu/kriz/cifar.html
15. Kurakin, A., Goodfellow, I.J., Bengio, S.: Adversarial examples in the physical world. In: ICLR (Workshop). OpenReview.net (2017)
16. LeCun, Y., Haffner, P., Bottou, L., Bengio, Y.: Object recognition with gradient-based learning. In: Forsyth, D.A., Mundy, J.L., Gesù, V.D., Cipolla, R. (eds.) Shape, Contour and Grouping in Computer Vision. Lecture Notes in Computer Science, Vol. 1681, p. 319. Springer, Heidelberg (1999). https://doi.org/10.1007/3-540-46805-6_19
17. Li, J., Schmidt, F.R., Kolter, J.Z.: Adversarial camera stickers: a physical camera-based attack on deep learning systems. In: ICML. Proceedings of Machine Learning Research, vol. 97, pp. 3896–3904. PMLR (2019)
18. Liu, Q., Li, P., Zhao, W., Cai, W., Yu, S., Leung, V.C.M.: A survey on security threats and defensive techniques of machine learning: a data driven view. IEEE Access 6, 12103–12117 (2018)
19. Madry, A., Makelov, A., Schmidt, L., Tsipras, D., Vladu, A.: Towards deep learning models resistant to adversarial attacks. In: 6th International Conference on Learning Representations, ICLR 2018, Vancouver, BC, Canada, April 30 - May 3, 2018, Conference Track Proceedings. OpenReview.net (2018)
20. Papernot, N., et al.: Technical report on the cleverhans v2.1.0 adversarial examples library. arXiv preprint arXiv:1610.00768 (2018)

21. Rauber, J., Brendel, W., Bethge, M.: Foolbox: a python toolbox to benchmark the robustness of machine learning models. In: Reliable Machine Learning in the Wild Workshop, 34th International Conference on Machine Learning (2017). http://arxiv.org/abs/1707.04131

22. Serban, A.C., Poll, E.: Adversarial examples - a complete characterisation of the phenomenon. arXiv preprint arXiv:1810.01185 (2018)

23. Szegedy, C., et al.: Intriguing properties of neural networks. In: ICLR (Poster) (2014)

24. Taori, R., Kamsetty, A., Chu, B., Vemuri, N.: Targeted adversarial examples for black box audio systems. In: IEEE Security and Privacy Workshops, SP Workshops 2019, San Francisco, CA, USA, pp. 15–20. IEEE (2019). https://doi.org/10.1109/SPW.2019.00016

25. Xiao, H., Rasul, K., Vollgraf, R.: Fashion-mnist: a novel image dataset for benchmarking machine learning algorithms. arXiv preprint arXiv:1708.07747 (2017)

26. Yakura, H., Sakuma, J.: Robust audio adversarial example for a physical attack. In: IJCAI, pp. 5334–5341. ijcai.org (2019). https://doi.org/10.24963/ijcai.2019/741

27. Zhao, Z., Dua, D., Singh, S.: Generating natural adversarial examples. In: ICLR (Poster). OpenReview.net (2018)

28. Zhu, J.Y., Park, T., Isola, P., Efros, A.A.: Unpaired image-to-image translation using cycle-consistent adversarial networks. In: Proceedings of the IEEE International Conference on Computer Vision, pp. 2223–2232 (2017)

Application of Graph Neural Network in Automatic Text Summarization

Rui Luo, Shan Zhao[✉], and Zhiping Cai

College of Computer, National University of Defense Technology, Changsha, China
{zhaoshan18,zpcai}@nudt.edu.cn

Abstract. The core of automatic text summarization technology is to extract the main content of input text information, remove redundant information in the text, and retain its core information to improve people's reading efficiency. However, a lot of research work on automatic text summarization still focus on modeling the explicit information of the text when extracting summaries, and the mining of hidden information is not strong enough. In previous work, researchers tried to introduce neural networks into automatic text summarization technology. The researchers found that automatic text summarization through neural networks can obtain hidden information in the text, but it is difficult for general neural networks to process long text information. In the past two years, graph neural networks (GNN) have begun to rise. Researchers have found that through the application of graph structure, long text input has been well processed, and can dig deeper into the hidden information in the text, so that the quality of the generated abstract has been significantly improved compared with previous studies. The introduction of GNN has also greatly promoted the transition of automatic text summarization from single-document summarization to multi-document summarization. And GNN has played an important role in promoting the development of automatic text summarization technology. In this article, we introduced the application of GNN in the field of automatic text summarization in recent years, summarized commonly used data sets and evaluation indicators, and summarized and prospected the development trend of automatic text summarization.

Keywords: GNN · Automatic text summarization · Extractive summarization · Generative summarization

1 Introduction

Currently people can get the information they want from the massive network data. Taking the text information as an example, the text information obtained from the network is numerous and may be repeated, so how to process the relevant text information more efficiently is a very important problem. Researchers are beginning to explore the use of programs to automate the processing of text information to generate concise content, known as automated text summaries.

© Springer Nature Singapore Pte Ltd. 2021
K. He et al. (Eds.): NCTCS 2020, CCIS 1352, pp. 123–138, 2021.
https://doi.org/10.1007/978-981-16-1877-2_9

By reading the summary of text information, people can grasp the main content of the text more quickly and efficiently, which can effectively improve reading efficiency. The essence of automatic text summarization is that the output text is much less than the input text on the premise of retaining the main information of the text.

There are two main ways of automatic text summarization:

(1) Extractive automatic summarization method, which extracts keywords and key sentences that already exist in the document, and then form a summary that is fluent in writing and retains most effective information.
(2) Generative automatic summarization method, which extracts the semantic information of the document by using neural network, and then regenerates a summary with smooth writing and retaining most effective information.

Since the generative automatic summarization method is limited by the slow development of information extraction technologies such as semantics, its effect is to some extent lagging behind the extractive automatic summarization. Therefore, in practical application, the more efficient extraction automatic summary method is the mainstream method at present.

The extractive automatic summarization method believes that the core content of a document can be summarized by a certain sentence or a combination of several sentences in the document. Therefore, the task of summarizing documents in extractive automatic summarization becomes to find the key sentences in the text. The easiest way to deal with extractive automatic summarization by computer is to sort. Sort based on the statistical results, calculate according to the potential information of the word, such as the position of the word, give each sentence a weight, and then sort the weights to select the most important sentence. Based on the statistical sorting method, the application of document information only stays at the surface information of words and sentences, and the potential information is not deeply explored, but this method is simple and easy to use. Sorting based on the graph structure, constructing a topological structure map of the document content, mining its potential information, and sorting words and sentences through topic models and other methods.

Generative automatic abstract mainly trains a large amount of text data through deep learning, deeply mines the potential information in the text, and generates a summary of the main content of the document by means of encoding and decoding. Generative automatic summarization can generate summaries for various texts. Influenced by the development of deep learning, the effect of current generative automatic summarization can be compared with that of abstract automatic summarization to a certain extent. However, due to the poor support of neural networks for long texts, automatic summarization is not suitable for comparing long texts, which is also a difficult point.

At present, most of the work on automatic summarization is still on single document summarization, and the research on multi document summarization will not have a great breakthrough until 2019. Especially after researchers began to pay more attention to GNN, the problem of long text dependence has been solved. According to different implementation methods, the development history of multi-document abstracts is briefly divided [18]. The first is extractive multi-document summarization. At that time, researchers mainly use graph sorting [30] and other methods to summarize multiple documents. The main content of this method is to sort the sentences by comparing the importance of different documents, and then select the most important sentences for combination. Next is document compression. Researchers delete and compress documents based on syntax or article structure to obtain multi-document summaries. During this period, researchers proposed methods for sentence compression and sentence fusion. Then there is the neural network-based approach, in which the researchers train the model by introducing neural networks. In the early days, since there was no publicly available large-scale data set in the field of multi-document summarization, researchers usually deal with the task of multi-document summarization on a model trained on single-document summarization. In this process, in 2018, researchers published a multi-document summary large-scale dataset Wikisum [20] containing 150W training samples. In 2019, researchers published a medium-scale news-oriented multi-document summary data set Multinews [6]. The last stage is to make targeted adjustments to the neural network used for multi-document summarization based on large-scale data sets.

Nowadays, deep learning is booming, with various kinds of neural networks emerging in an endless stream, and the application of neural networks can be seen in various fields. Among them, the research and application of neural network in computer vision and natural language processing are more in-depth. With the support of neural network, automatic text summary, as a downstream task of natural language being processed, has been developed rapidly. Since 2019, GNN has had a great influence in the field of computer vision, some researchers have introduced GNN into natural language processing. After a period of research, whether it is extracted automatic text summarization or generated automatic text summarization, GNN can have a good influence on it, which greatly promotes the development of automatic text summarization.

The contributions made by our paper are summarized as follows:

(1) We provide an overview of automatic text summarization based on GNN. We describe in detail representative models.
(2) We summarized the resources on GNN-based automatic text summarization, including benchmark data sets and evaluation indicators.
(3) We analyzed the limitations of existing work and proposed some possible future research directions.

In this paper, the second part introduces the development of GNN and the related concepts of graph convolution network. The third part summarizes the related applications of current GNN in the field of automatic text summarization, introduces several representative models. The fourth part summarizes the widely used benchmark data sets and evaluation indicators. The fifth part discusses current challenges and suggests future research directions. The sixth part summarizes the thesis.

2   Research Background

2.1   The Development History of GNN

Since the concept of neural network was put forward, neural network has made great progress. One after another neural networks with significant effects have been proposed by researchers and continuously improved, including convolutional neural networks (CNN), recurrent neural networks (RNN), deep neural networks (DNN), adversarial neural networks (GAN), and so on. Among them, Micheli et al. [22] proposed a graph neural network in 2009 to process data with graph structure. However, the original graphical neural network has not received much attention due to its limitations in representation ability and training efficiency.

CNN is undoubtedly the most widely used neural network in computer vision at present. The achievements made by CNN have opened a prelude to the further development of deep learning. However, CNN also has a defect that it cannot handle data with graph structure. Therefore, it is necessary to find another way to deal with the data with graph structure. At the same time, with the development of Graph Embedding technology, researchers put forward Deep Walk, LINE, SDNE and other methods to be applied to network learning representation. These methods can effectively process the data of graph structure through graph embedding learning the feature representation of nodes, edges, or subgraphs in the graph, which provides important inspiration for the proposal of graph neural networks. However, graph embedding technology cannot achieve the optimal situation in computing large-scale and more complex graphs.

With the increasing number of people studying convolution network, some researchers put forward some new concepts about convolution of graph structure data in previous years, put forward convolution network which can be applied to graph structure. These methods are collectively called graph convolution network (GCN) [13]. Bruna et al. (2013) [2] were inspired by the spectrogram theory, proposed a neural network for graph convolution on this basis. Since then, researchers have continued to improve and extend graph convolutional networks in Bruna et al. Because the spectrum method convolves the whole graph, this method can't deal with large-scale graph structure, so researchers began to change their thinking and focus on the graph convolution network based on space. The graph convolution network based on space collects the information of adjacent nodes through the topological structure of graph, so it can convolve some nodes of graph. Combined with sampling strategy, the calculation process can be simplified, and the calculation objects can be reduced from the whole graph to some nodes, thus improving the operation efficiency.

In the past two years, researchers have refocused their attention on graph neural networks. As more and more researchers join, new structures of graph neural networks are constantly being proposed. These new structures of neural networks are relative to the original graph. The representation ability and training efficiency of neural networks have been greatly improved. The graph attention network (GAT) [26] improves the information collection of the graph structure in the model by introducing the attention mechanism, and further mines the hidden information of the graph structure.

2.2 The Basics of GCN

At present, most of the automatic text summarization involving GNN is developed and researched based on graph convolutional network, so it is very necessary to understand the principle of graph convolutional network. Graph convolutional networks can be divided into two types, the graph convolution network based on spectrum and the graph convolution network based on space.

The GCN Based on Spectrum. In the graph convolution network based on spectrum, the graph can be represented by a regularized Laplacian matrix, where the graph is an undirected graph. For the feature vector x of each node in the graph, the Fourier transform of the graph is performed on it, and it is mapped into a standard orthogonal space. The formulas involved are shown in Table 1. For input x, its graph convolution can be expressed as:

$$x * Gg = F^{-1}(F(x) \odot F(g)) = U\left(U^T x \odot U^T x\right) \tag{1}$$

$\odot$ means Hadamard product.

Therefore, the choice of G is very important for graph convolution network based on spectrum. Graph convolution algorithms based on spectrum include Spectral CNN, Chebyshev Spectral CNN, First order of ChebNet, adaptive graph convolution network and so on.

Table 1. Formulas involved in graph convolutional networks

Formula	Description
$L = D - W$	The definition of Laplace matrix, where D is the degree matrix of the graph, which is a diagonal matrix, and W is the adjacency matrix of the graph
$L = l - D^{-\frac{1}{2}}WD^{-\frac{1}{2}}$	Representation of the standard Laplace matrix
$L = U \wedge U^T$	The standard Laplacian matrix is a real symmetric positive semi-definite matrix, so it can be eigen-decomposed. This formula is the eigen-decomposition of the standard Laplacian matrix. Where L is a diagonal matrix, the values on the diagonal are the sorted eigenvalues, and U is the eigenvector matrix corresponding to the eigenvalues
$F(x) = U^T x$	Fourier transform
$F^{-1}(x) = U\hat{x}$	Inverse Fourier transform

The GCN Based on Space. The emergence of graph convolution network based on space is affected by the convolutional operation of convolutional neural networks on images. The difference is that the graph convolution network based on space define graph convolutions based on the spatial relationship of nodes. For graph convolution, a new representation of the node can be obtained by aggregating the nodes in the graph with their neighbor nodes.

Like convolution kernel in traditional deep learning, in space-based graph convolution, graph convolution operator is defined as:

$$h_i^{(l+1)} = \sigma \left(\sum_{j \in N(i)} \frac{1}{c_{ij}} W^{(l)} h_j^{(l)} \right) \tag{2}$$

$h_i^{(l+1)}$ represents the feature information of node i at the $1 + 1$ layer; c_{ij} represents the normalization factor, such as the degree of the node; $W^{(l)}$ indicates the direct weight of the node; $h_j^{(l)}$ Indicates the feature information of node j at the l layer.

When the graph convolutional network performs convolution operations, first each node transfers its own feature information to all neighbor nodes, and then each node integrates the collected feature information to obtain feature information of a local graph structure, and then the activation function is added, and the characteristic information of the node is nonlinearly transformed to enhance the expressive ability of the model. Therefore, the key to the graph convolutional network is to learn a function that can integrate the characteristic information of the current node and its neighbor nodes.

3   Research Status and Hot Spots

3.1   The Preliminary Application of GNN in Automatic Text Summarization

Before neural networks became a research hotspot, extractive summaries occupied a major position in the field of automatic summarization. Using neural networks, the effect of generative summaries has begun to match and even surpass that of extractive summaries. The rise of graph neural networks has set off a big wave in the field of automatic summarization. In 2018, Patrick Fernandes et al. [8] proposed a hybrid model, in which the standard sequence encoder is a graph neural network, and successfully applied the graph neural network to sequence data with weaker structure and made extensive evaluation of three summary tasks in the literature. Based on the graph neural network, both extractive automatic summarization and generative automatic summarization have been further improved.

Earlier, most GNNs dealt with homogeneous graph structures, and the node representation of such graphs was learned through models. However, when the structure of the graph is uncertain or the processing object is a heterogeneous graph, the general GNN will have many problems during training. Seongjun Yun et al. [31] proposed a new graph transformation network – Graph

Transformer Networks (GTN) in 2019, which can learn effective node representation on graphs by identifying useful meta-paths and multi-hop connections.
This effectively solves the problem of learning node representation on uncertain
or heterogeneous graphs. This makes GTN quite comfortable with a wide variety
of documents when it comes to auto-summarization.

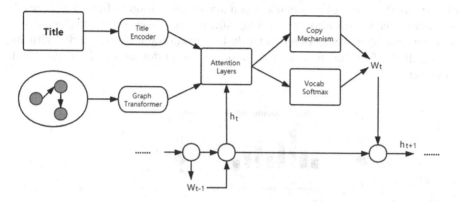

Fig. 1. GraphWriter model overview

Automatic summarization requires the system to identify important entities
in the input and their relationships, and to filter out redundant information. And
one of the key issues is to extract valid information across multiple sentences.
Through structured representation, effective information can be extracted across
multiple sentences, but extracting such effective information manually will consume a lot of human resources. Rik Koncel et al. [14] used graph transformers to
extract features from the graph, and established an end-to-end trainable Graph-
Writer model, which can generate a complete paragraph through the knowledge
graph. Figure 1 is the architecture diagram of the GraphWriter model. Experiments show that this graph-to-text end-to-end system can generate document
abstracts well.

3.2   GNN with Attention Mechanism

In the task of automatic summarization, the relationship between sentences
can be modeled in two ways, the first is a sequence model represented by
RNN(LSTM), and the second is a Graph-centered model. The sequence model
in the first way is difficult to capture the long-distance dependence of sentence
level when dealing with long texts, and the short-distance information of context between adjacent sentences has great influence on modeling. In contrast,
the model based on graph structure pays more attention to global information,
which is undoubtedly very important for automatic summarization task.

At present, when extracting text information, much research only captures
the explicit information of text, such as sentence position, word nature, entity

name and so on. Haiyang Xu et al. [28] connected the analytic tree of sentences in the document with graphs and used stacked GCN to learn the syntactic representation of documents. The model architecture is shown in the Fig. 2. From sentence position, noun to syntactic representation, semantic structure, etc., the text information that can be extracted by graph neural network is becoming more and more abundant, including both explicit information and hidden information of text, so the effect of automatic summary of generation is further enhanced. Attention mechanism also played a significant role. In previous studies, attention mechanism played a dominant role in sequence generation model with its outstanding effect and was used to improve the performance of abstract text summarization.

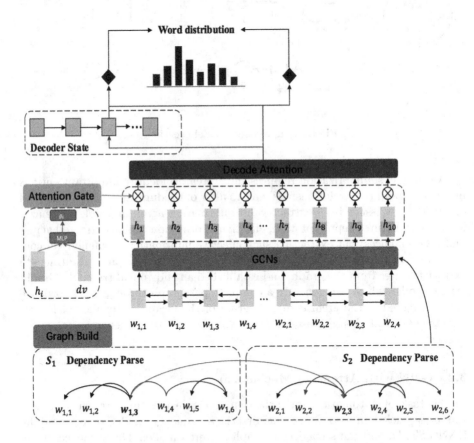

Fig. 2. Overall architecture of the proposed model [28]

In previous work, attention mechanism is usually combined with CNN to form a codec structure, which has achieved impressive results in abstract abstracts of news articles. However, these methods do not consider long-term dependencies in document sentences. The two papers published by Harvard NLP are both text abstracts about SEQ2SEQ, which introduce attention mechanism to optimize encoder based on SEQ2SEQ to improve the effect of text abstracts. By combining the attention mechanism and SEQ2SEQ, a good optimization result is obtained, which provides a good reference for the follow-up study. Facts have proved that the graph neural network with the attention mechanism does have a significant improvement in performance [15,24].

3.3   Improved GNN for Encoding Objects

In previous work, researchers often used GCN for text generation, and proposed many implementation methods. Among them, some methods are based on global nodes (assuming that each node is connected to other nodes in the graph), but this method does not consider the structure of the graph. Other methods are based on local nodes (only a part of the nodes is connected) for calculation, but these methods do not make good use of the global information of the graph. For this reason, Leonardo et al. [23] proposed four models that can combine global nodes and local nodes. The model architecture is shown in the Fig. 3. By combining the two, it can be ensured that the model can model the entire graph structure while acquiring the local graph structure information. The method is tested on the datasets AGENDA and WebNLG, and the results are significantly improved compared with the previous method, which proves the effectiveness of the method.

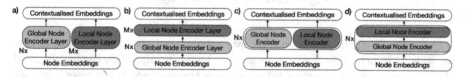

Fig. 3. Schematic diagram of four models, which are called PGE, CGE, PGE-LW and CGE-LW respectively from left to right [23].

Leonardo et al. proposed two methods to combine local graph structure with global graph structure, and finally formed four models. These two methods are parallel coding and cascade coding. Parallel coding firstly encodes the local graph structure and the global graph structure respectively, and then splices them together. Cascade coding is to carry out global graph coding first, and then take the feature vector of global graph coding as the input of local graph coding, and then get the result. In global graph coding, the features of all other nodes are weighted and averaged first, and then multiple vectors are spliced in combination with multi-head attention mechanism to get the final vector. Local graph coding evaluates the features of all adjacent nodes by weight, and others are consistent with global graph coding. At the same time, these two methods are divided into hierarchical and non-hierarchical. Non-hierarchy is carried out independently, and hierarchy is carried out in units of layers.

3.4 Improved GNN for Graph Structure

A long time ago, graph structures represented by LexRank [5] and TextRank were used in extractive automatic summarization tasks. LexRank and TextRank take sentences as the basic unit as the nodes of the graph structure, and the edges of the graph structure are the similarity between sentences as shown in Fig. 4. After sampling the information of adjacent nodes, the importance is ranked by unsupervised iteration. The most important sentences were selected from the results and regrouped as the final summary. However, with this kind of graph, it is difficult to determine the appropriate range when selecting the similarity between sentences [3].

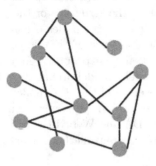

Fig. 4. LexRank structure diagram

Before 2020, when extractive automatic summarization extracts text features, the position information of the sentence has a greater impact on the feature, and relatively speaking, the content information of the sentence has a general impact on the feature. For example, deleting modifiers, clauses, etc. in the input sentence will not have a greater impact. Danqing Wang et al. [27] try to further introduce word nodes based on sentences to enrich the graph structure, further clarify the relationship between sentences through word nodes, and model heterogeneous graphs to obtain abstracts. By introducing word nodes and iteratively updating word nodes and sentence nodes in modeling, the role of words in sentence representation can be strengthened. The relationship between sentences can be better established through word nodes, and the relationship between multiple documents can also be established through word nodes. Thus, the HeterSum-Graph model can be easily migrated from a single document summary to a multi-document summary task.

3.5 GNN with Pre-trained Model

In 2018, the pre-training model Bert [4] attracted wide attention once it was proposed. Bert model has a significant impact on NLP tasks, which makes many researchers turn their attention to the field of pre-training. After that, researchers have applied Bert model and graph neural network to automatic text summarization, but the effect is not as obvious as expected. Because Bert model pre-trains sentence pairs, not the whole document, it is not particularly effective in the task of abstract generation for long texts. However, Jiacheng Xu et al. [29] proposed a DISCOBERT model, which is a DISCOBERT based abstraction model for the sensory neurons of a discourse. The model architecture is shown in the Fig. 5. The author captures the long-term dependency relationship in the document through a discourse graph based on Rhetorical Structure Theory (RST) tree and co-reference relationship, and then encodes it using the graph convolutional network. To better use the Bert model, the author did not choose the sentence as the minimum selection unit, but the basic discourse unit (EDU), which is a clause phrase unit derived from RST. By operating on EDU, the model is better able to discard redundant information from the sentence, saving the main underlying information in the sentence, and thus providing a more concise and informative summary.

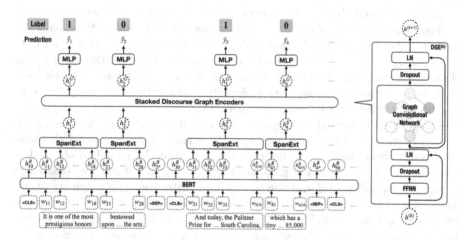

Fig. 5. (Left) DISCOBERT's model architecture. The stackable speech graph encoder contains k stackable DGE blocks. (Right) Architecture of each speech graph encoder (DGE) block [29].

3.6 The Application of Graph Neural Network in Multi-text Abstract

In the field of automatic text summarization, single-document summarization has always been the focus of researchers' attention, until recently, the GNN has made great progress in the field of automatic summarization, researchers began to pay attention to multi-document summarization. For multi-document summary tasks (MDS), how to improve the long text representation capability of neural networks is a huge challenge [7,32]. Unlike a single document summary, the information gap between different documents in a multi-document summary task is large, or even completely unrelated, so it requires a stronger ability of summarization and induction.

Therefore, how to model the relationship between documents is a very important issue in MDS. Only by successfully extracting the relationship between documents can key information and redundant information be effectively distinguished, which will directly affect the final summary [1]. Yang Liu et al. [21] express cross-document relationship through a attention mechanism, which allows information to be shared across multiple documents, thus effectively extracting the relationship between documents. By constructing the graph structure, the relationship between documents can also be well represented [16]. Wei Li et al. [18] proposed to apply three graph structures: semantic graph, discourse graph, and topic graph to the task of generating multi-document summarization. Prior to this, these three graphs were commonly used in extractive multi-document summarization tasks [17]. Experimental results show that the quality of the result can be significantly improved by introducing these graph structures.

Similarly, Hanqi Jin [12] and others regard documents, sentences, and words as three semantic units of different granularities [19] and connect these semantic units in a three-granularity hierarchical relationship diagram, as shown in Fig. 6. A new multi-granularity interaction network is proposed to facilitate the learning of monitoring all granularity representations. In the coding stage, the attention mechanism is used to encode the relationship between the same semantic granularity and the hierarchical relationship between different semantic granularity. Fusion gates are then used to integrate various relationships to update the semantic representation. In the decoding stage, the extracted and generated abstracts are trained simultaneously to improve the ability of the model to obtain key information under different situations. It has become a hotspot for researchers to capture the hidden information of text by constructing the graph structure. According to the corresponding experimental results, the graph structure can effectively capture the hidden information and provide strong support for the generation of the final abstract, no matter it is a single document summary or a multi-document summary.

4   Data Sets and Evaluation Criteria

We have made statistics on the common data sets in the previous work and listed the publicly available data sets in alphabetical order, as shown in Table 2. The size and complexity of data sets can be adjusted according to different practical problems. BigPatent [25] includes 1.3 million records of American patent documents and written abstracts of human beings. DUC2004 uses news lines/paper documents from TDT and TREC sets, which is an English data set that can be used for testing. Gigawords is an English sentence abstract data set. By extracting the first sentence and title from news reports, a source and abstract pair (i.e., the first sentence and title) is formed. The LCSTS [11] consists of more than 2 million authentic Chinese essays, each of which is provided with a summary,

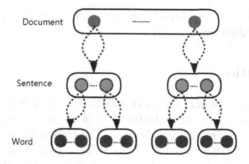

Fig. 6. Three-granularity hierarchy diagram

and the data is collected from Sina Weibo [10]. Newsroom [9] is a large data set for training and evaluating summary system. It contains 1.3 million articles and abstracts, which are written by authors and editors in the newsrooms of 38 major publications.

Table 2. Data set and its download address

Data set name	Download link
BigPatent	https://evasharma.github.io/bigpatent/
CNN/Daily Mail	https://github.com/deepmind/rc-data/
DUC 2004	https://duc.nist.gov/duc2004/tasks.html
Gigaword	https://drive.google.com/open? id=0B6N7tANPyVeBNmlSX19Ld2xDU1E
LCSTS	http://icrc.hitsz.edu.cn/Article/show/139.html
Multinews	https://github.com/Alex-Fabbri/Multi-News
Newsroom	https://summari.es/download/
Wikisum	https://github.com/tensorflow/tensor2tensor/tree/ 5acf4a44cc2cbe91cd788734075376af0f8dd3f4/ tensor2tensor/data_generators/wikisum

At the same time, we also summarize some commonly used model evaluation standards, such as Rouge, MoverScore, BertScore and so on. Rouge compares the abstract content generated by the model with a set of reference abstracts prepared in advance and obtains the corresponding scores. Through the scores obtained, the similarity between the automatically generated abstracts and the real results is determined. Rouge can be roughly divided into four types: ROUGE-N, ROUGEL, ROUGE-W and Rouge-s. BertScore obtains the sentence context representation through Bert model, and then gets a score according to the artificial design calculation logic, which is the similarity of sentences. MoverScore improves and innovates based on BertScore. Rouge is a classic evaluation index, while BertScore and MoverScore are more scientific and effective as the newly proposed evaluation indexes.

5   Future Outlook

In the past few years, graph neural networks have been used on a large scale and have received extensive attention. In the field of automatic text summarization, graph neural networks are also widely used. The core problem of automatic text summarization task is that it is difficult to accurately capture key context information from long text sequences through existing model conditions. However, the application of graph neural network effectively solves this problem, and the hidden information in long text can be captured by constructing graph structure, which is reflected in recent research. Another mechanism that can help the model capture hidden information is attention mechanism. By adding attention mechanism to the model, the ability of the model to deal with long texts can also be greatly enhanced. In practical experiments, a well-constructed graph has better ability to capture grammatical or semantic relations than attention mechanism, but the combination of the two maximizes the ability to obtain comprehensive and accurate context-dependent vectors. Therefore, a good graph structure combined with an attention mechanism can become an important research direction to promote the development of automatic summarization.

Some existing research improve the training effect by introducing the pre-training language model. However, because the pre-training model is to pre-train sentence pairs, it cannot adapt well to the information extraction of documents in automatic text summarization. From the experimental results, there is only a little improvement effect. But this can still provide researchers with certain ideas. They can start with the pre-training model and improve the automatic text summarization task to make the pre-training model more suitable for use on the automatic text summarization task, thereby further improving the quality of summary generation.

In addition, early researchers studied extractive abstracts and generative abstracts separately. Now some researchers have noticed that the quality of long text abstracts can be improved by combining extractive abstracts and generative abstracts. No matter from which point of view, automatic text summarization is always developing, and the quality of the generated summaries will be higher and higher.

6   Summary

In this article, we have provided a detailed overview of automatic text abstractions based on GNN. We start with the research background of the graph neural network and introduce the development process and related concepts of the related fields. Then we also introduce the current application of graph neural network in the field of automatic text summarization and list several representative models. We also summarize some widely used benchmark data sets and assessment indicators. Finally, we discussed current challenges and suggested directions for future research.

References

1. Barzilay, R., McKeown, K.R., Elhadad, M.: Information fusion in the context of multi-document summarization. In: Proceedings of the 37th Annual Meeting of the Association for Computational Linguistics, pp. 550–557. Association for Computational Linguistics, College Park (June 1999). https://doi.org/10.3115/1034678. 1034760, https://www.aclweb.org/anthology/P99-1071

2. Bruna, J., Zaremba, W., Szlam, A., LeCun, Y.: Spectral networks and locally connected networks on graphs. CoRR abs/1312.6203 (2014)

3. Chen, Y.C., Bansal, M.: Fast abstractive summarization with reinforce-selected sentence rewriting. ArXiv abs/1805.11080 (2018)

4. Devlin, J., Chang, M.W., Lee, K., Toutanova, K.: Bert: pre-training of deep bidirectional transformers for language understanding. In: NAACL-HLT (2019)

5. Erkan, G., Radev, D.R.: LexRank: graph-based lexical centrality as salience in text summarization. J. Artif. Intell. Res. **22**, 457–479 (2004)

6. Fabbri, A.R., Li, I., She, T., Li, S., Radev, D.R.: Multi-news: a large-scale multi-document summarization dataset and abstractive hierarchical model. ArXiv abs/1906.01749 (2019)

7. Fan, A., Gardent, C., Braud, C., Bordes, A.: Using local knowledge graph construction to scale seq2seq models to multi-document inputs. In: EMNLP/IJCNLP (2019)

8. Fernandes, P., Allamanis, M., Brockschmidt, M.: Structured neural summarization. ArXiv abs/1811.01824 (2019)

9. Grusky, M., Naaman, M., Artzi, Y.: Newsroom: a dataset of 1.3 million summaries with diverse extractive strategies. ArXiv abs/1804.11283 (2018)

10. Hermann, K., et al.: Teaching machines to read and comprehend. In: NIPS (2015)

11. Hu, B., Chen, Q., Zhu, F.: LCSTS: a large scale Chinese short text summarization dataset. In: EMNLP (2015)

12. Jin, H., Wang, T., Wan, X.: Multi-granularity interaction network for extractive and abstractive multi-document summarization. In: Proceedings of the 58th Annual Meeting of the Association for Computational Linguistics, pp. 6244–6254. Association for Computational Linguistics (July 2020). https://doi.org/10.18653/ v1/2020.acl-main.556, https://www.aclweb.org/anthology/2020.acl-main.556

13. Kipf, T., Welling, M.: Semi-supervised classification with graph convolutional networks. ArXiv abs/1609.02907 (2017)

14. Koncel-Kedziorski, R., Bekal, D., Luan, Y., Lapata, M., Hajishirzi, H.: Text generation from knowledge graphs with graph transformers. In: NAACL-HLT (2019)

15. Krantz, J., Kalita, J.: Abstractive summarization using attentive neural techniques. ArXiv abs/1810.08838 (2018)

16. Li, P., Bing, L., Wei, Z., Lam, W.: Salience estimation with multi-attention learning for abstractive text summarization. ArXiv abs/2004.03589 (2020)

17. Li, W., Zhuge, H.: Abstractive multi-document summarization based on semantic link network. IEEE Trans. Knowl. Data Eng. **33**(1), 43–54 (2021). https://doi.org/ 10.1109/TKDE.2019.2922957

18. Li, W., Xiao, X., Liu, J., Wu, H., Wang, H., Du, J.: Leveraging graph to improve abstractive multi-document summarization. ArXiv abs/2005.10043 (2020)

19. Li, Z., Wu, W., Li, S.: Composing elementary discourse units in abstractive summarization. In: Proceedings of the 58th Annual Meeting of the Association for Computational Linguistics, pp. 6191–6196. Association for Computational Linguistics (July 2020). https://doi.org/10.18653/v1/2020.acl-main.551, https://www.aclweb.org/anthology/2020.acl-main.551

20. Liu, P.J., et al.: Generating Wikipedia by summarizing long sequences. ArXiv abs/1801.10198 (2018)

21. Liu, Y., Lapata, M.: Hierarchical transformers for multi-document summarization. In: ACL (2019)

22. Micheli, A.: Neural network for graphs: a contextual constructive approach. IEEE Trans. Neural Netw. **20**(3), 498–511 (2009). https://doi.org/10.1109/TNN.2008.2010350

23. Ribeiro, L.F.R., Zhang, Y., Gardent, C., Gurevych, I.: Modeling global and local node contexts for text generation from knowledge graphs. Trans. Assoc. Comput. Linguist. **8**, 589–604 (2020)

24. Rush, A.M., Chopra, S., Weston, J.: A neural attention model for abstractive sentence summarization. ArXiv abs/1509.00685 (2015)

25. Sharma, E., Li, C., Wang, L.: BigPatent: a large-scale dataset for abstractive and coherent summarization. In: ACL (2019)

26. Velickovic, P., Cucurull, G., Casanova, A., Romero, A., Liò, P., Bengio, Y.: Graph attention networks. ArXiv abs/1710.10903 (2018)

27. Wang, D., Liu, P., Zheng, Y., Qiu, X., Huang, X.: Heterogeneous graph neural networks for extractive document summarization. ArXiv abs/2004.12393 (2020)

28. Xu, H., Wang, Y., Han, K., Ma, B., Chen, J., Li, X.: Selective attention encoders by syntactic graph convolutional networks for document summarization. In: ICASSP 2020 - 2020 IEEE International Conference on Acoustics, Speech and Signal Processing (ICASSP), pp. 8219–8223 (2020)

29. Xu, J., Gan, Z., Cheng, Y., Liu, J.: Discourse-aware neural extractive model for text summarization. ArXiv abs/1910.14142 (2019)

30. Yin, Y., Song, L., Su, J., Zeng, J., Zhou, C., Luo, J.: Graph-based neural sentence ordering. In: Proceedings of the Twenty-Eighth International Joint Conference on Artificial Intelligence, IJCAI-19, pp. 5387–5393. International Joint Conferences on Artificial Intelligence Organization (July 2019). https://doi.org/10.24963/ijcai.2019/748

31. Yun, S., Jeong, M., Kim, R., Kang, J., Kim, H.: Graph transformer networks. In: NeurIPS (2019)

32. Zhang, J., Tan, J., Wan, X.: Towards a neural network approach to abstractive multi-document summarization. ArXiv abs/1804.09010 (2018)

Head Detection Method for Indoor Scene

Zhi Li[1(✉)], Yong Li[2], and Xipeng Wang[1]

[1] Engineering University of PAP, Postgraduate Brigade, Xian, China
[2] College of Information Engineering, Engineering University of PAP, Xian, China

Abstract. This paper proposes a trainable fully convolutional, end-to-end crowd head detector model, which is a single convolutional network architecture, applying for target bounding box prediction and classification tasks in images. Due to the network structure, this model is very lightweight in terms of parameters, and needs less calculation time and memory space. Experiments showed that our model had a better overall average precision than some head detection benchmark algorithms, and can achieve this goal by effective receiving domains of model's network to choose different anchor sizes. By testing on the classic head detection datasets, it is concluded that our model can achieve a 0.69 AP result on the challenging head detection Brainwash dataset, which is comparable to those head detection benchmark methods. At the same time, the best result was achieved on the Holly-woodHeads dataset. In addition to accuracy, our model also has the advantages of less parameters, less computing time, and less storage space requirements. It has a certain reference significance for making an intelligent monitoring system.

Keywords: Head detection · Target recognition · Crowd count · CNN · Effective receiving domain

1 Introduction

Head detection in crowded scenes is an important issue in the field of computer vision. It has a wide range of applications, such as intelligent monitoring. Crowd counting based on head detection methods, comparable to previous crowd counting techniques based on density map methods [1–3], can give more reliable and accurate results. It is not always possible to get the correct location information and the final number of people detected with the density map method, which will cause the detection results to become unreliable and wrong. And non-detection targets are also judged as detection targets. In the past, researchers have also proposed several methods for human head detection, using different local context information [4–9]. As [6] proposed Human body segmentation to locate the head of the person in the image. [10] trained a complex classifier based on the cascade of improved integral features order to implement analysis-based head detection, Marin et al. [9] first locate the upper part of the human body, then trained the DPM [11] to detect the head in the upper part of the human body that has been positioned. This method is to understand how "location" and "detection" are integrated in the video

© Springer Nature Singapore Pte Ltd. 2021
K. He et al. (Eds.): NCTCS 2020, CCIS 1352, pp. 139–146, 2021.
https://doi.org/10.1007/978-981-16-1877-2_10

scenes. In recent years, many methods of head detector [7, 8] based on deep learning are proposed. One of the best performing head detection models [7] extends the R-CNN detector for head detection, using two contextual information prompts. The first type of information prompt makes the global CNN model use the relationship between people and target scenes. The second type of information prompt uses structured output to model the relationship between the detected objects. These two types of prompt information are combined in a joint R-CNN framework. There are two methods for head detection [7, 9], both are used in the following scenarios: the average head size is larger and the average head count is less. When these models are tried for head detection in crowded scenes, it may not be applicable. Because the head detection situation in this scene may be completely opposite, such as a smaller average head size and more average head count. Without considering the combination of large size and density changes in previous model design, the entire task of head detection in crowded scenes has become an urgent problem that needs to be solved. The whole work of this paper is closely related to [8], [8] proposes a method in crowded scenes, an anchorless detection framework for head detection. This framework has RNN to use the information in the deep representation to make predictions. Our paper proposes a structural framework for head detection, which is especially suitable for head detection in indoor crowded scenes. Our model is based on some conventional anchors, these anchors are based on the detection pipeline in the Faster-RCNN [12] method, which are obtained by regularly tiling a set of prediction-boxes with different sizes in the pictures. However, unlike Faster-RCNN, the two-stage pipeline processing mode, our model is different. The network structure of this article has only one full convolutional network that can perform classification and bounding-box prediction at the same time.

In this work, we have some contributions as following: (1) Propose a fully, end-to-end trainable convolutional head detection model, which is based on effective receiving domain in network, can systematically select the anchor size and achieve great detection results in crowded scenes; (2) Our model needs low memory requirements and calculation time, and can be considered to install on an embedded device. Experiments have shown that our model has achieved better results compared with other architectures, and the test results on the challenging head detection datasets are better than some previous benchmark algorithms.

2 Network Structure

2.1 Model Structure

Deep network architectures like [13–15] have the ability to learn "common features", which are very useful in various practical tasks. We first uses the pre-trained VGG-16 [15] model as the basic of our model Part. Remove the fully connected layer network behind VGG-16, and then use the remaining weights as the starting point for new training (see Fig. 1).

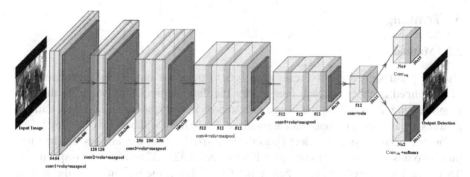

Fig. 1. Network structure

After the pre-trained VGG-16 model, we added 3 new convolutional layers: (1) Use the first convolutional layer to slide on the entire feature map to detect all positions on; (2) Use the second convolutional layer to make regression prediction of the positioning coordinates of the human head in the images; (3) The third convolutional layer combines softmax for head classification to predict the score of the detected human head targets. The second and third convolution layer both use 1×1 convolution kernels. After the output of the regression network is transformed by the bounding box, the size and displacement predictions obtained are converted into spatial coordinates. The number of convolutional layer kernels used for classification and regression depends on each pixel position's (N) number of anchors.

2.2 Design of Anchor Size

These anchors are obtained by placing pre-defined proportional detection frames at regular intervals in the image. These anchors represent the size and position offset predicted by the network in the target detection system [12]. We select "effective receiving domain" instead of "theoretical receiving domain" to choose the anchor size. It is pointed out in [16] that each unit in the CNN network has two types of receiving domain, which are theoretical receiving domain and effective receiving domain. Experiments prove that theoretical receiving domain has only a few central pixels in the field are responsible for triggering the neurons in the final output. The central area of the theoretical receiving domain is the effective receiving domain. In the Conv5 layer of the VGG-16 network, the theoretical receiving domain size is 228 pixels. Through experiments, we found it is more suitable that anchors with scale ratios of 2 and 4 in practice use, which represent the anchor size 32×32 and 64×64 respectively. The calculation method of anchor size is: "convolutional layer step length" × "azimuth ratio" × "anchor scale ratio", for example, the step size of the Conv5 layer is 16. The anchor size is scaled down to 3.5 times of the theoretical receiving domain to match the effective receiving domain size. The anchor size designed in this way helps to obtain good detection results, especially in crowded indoor scenes with smaller human head size. This conclusion can be drawn from experiment results.

3 Training

3.1 Method

Use the pre-designed anchor scale ratio sets to train our model, and generate a set of predefined anchors for each image after the data preprocessing stage. In the data preprocessing stage, adjust the image size to 640 × 480 resolution, and then training pictures are averagely cropped and normalized. Since there will be 4 max-pooling layers before the Conv5 layer, we select 16 as the step size to cover the anchors. At the same time, two sets of anchors are selected, whose sizes are 32 × 32 and 64 × 64 respectively. There are 4 max-pooling layers before the Conv5 layer, the resolution of the feature map obtained at the Conv5 layer is 40 × 30 (the resolution of the input image is 640 × 480). For each picture, 600 anchor points will be obtained (20 × 15 × 2), and for each anchor point, the network will predict 4 regression coordinates through the $Conv_{reg}$ layer, and predict the probability scores of two detected heads through the $Conv_{cls}$ layer. Regression coordinates use the form of a scale, and move the anchor point to locate the actual head position in the target scene, regardless of the situation where the anchor point exceeds the image boundary. In order to label the anchors and train the network, the whole training process must obey the following two rules: (1) regard anchors with IoU (Intersection over Union) ≥ 0.7 as positive labels; (2) The anchor with the largest IoU is regarded as a positive label. This work follows these two rules, because in some cases, the first rule may not be able to mark some anchors as positive labels. Anchors with IoU < 0.3 will be regarded as negative labels, and anchors that do not meet the above two rules will be ignored and not be included in the training process. During the training, only one picture is processed at a time. From the anchors that have been assigned positive and negative labels, a total of 32 positive and negative label anchors are selected at a ratio of 1:1. Throughout the experiment, the following conditions remained unchanged: Non-Maximum Suppression (NMS) is used in the test process instead of the training process. NMS is a post-processing technique used to filter out high overlapping detection results. And NMS stipulates: if there are two IoU results are more than the preset threshold, then these two detections will be suppressed to one. The NMS threshold is set to 0.3, and the following evaluation indicators are used: If the predicted bounding box IoU of the real image is more than or equal to 0.5 [20], it is considered to be the correct head.

3.2 Loss Function

The loss function used for model training is the multi-task loss function, similar to the function defined in RPN [12] training. Its definition is shown in (1).

$$L(\{p_i\}, \{t_i\}) = \frac{1}{N_{cls}}\sum_i A + \frac{1}{N_{reg}}\sum_i B \tag{1}$$

$$A = L_{cls}(p_i, p_i^*) \tag{2}$$

$$B = p_i^* L_{reg}(t_i, t_i^*) \tag{3}$$

where i represents the selected index of the anchor, the range is $[0-31]$. p_i is the number of heads contained in the anchor i, p_i^* represents the predicted probability of the true label (1 or 0). t_i is the parameterized coordinates of the predicted bounding box (including scale ratio and translation size), and t_i^* is the true value of the parameterized coordinates. These parameterized coordinates are defined in the same way as [17]. L_{cls} means classification loss, L_{reg} means regression loss. L_{cls} is at all anchors, but L_{reg} only calculates the positive label anchors, so the p_i^* in front of the equation L_{reg} needs more attention. L_{cls} uses cross-entropy loss, and L_{reg} uses the smooth-L1 loss defined in [17]. These two losses are normalized by N_{reg} and N_{cls}, which represent the number of samples for classification prediction and regression prediction, respectively.

3.3 Hyper Parameter

The basic model is initialized by the pre-trained VGG-16 model. VGG16 is trained on the ImageNet dataset [18]. All the layers of the VGG-16 model and the new network layer need to be retrained. The new network layer uses the random weights sampled with standard normal distribution ($\mu = 0$ and $\sigma = 0.01$) to initialize. The weight attenuation is set to 0.0005, and the entire model is fine-tuned by the SGD optimizer. During the training, the learning rate is set to 0.001, and the model will be trained for 30 cycles (about 320K iterations). After 10 training cycles are completed, the learning rate is attenuated by a ratio of 0.1. The whole training process is carried out in the PyTorch framework [19].

4 Experiment

In order to fully evaluate the model proposed in this article, we use the training set (Brainwash, Hollywoodheads) of the corresponding dataset to train the model, and then evaluate the performance of the model on the test set. Table 1 shows the comparison of different benchmark methods with our model's performance on Brainwash dataset. The Brainwash dataset [8] was captured in crowded scenes with many human heads, and the average head count reached 7.89.

4.1 Result on Brainwash Dataset

We use the test set containing 484 images on Brainwash dataset to evaluate our detection model, and consider comparing the same benchmark methods with our method. These benchmarks have been used in [8], that is, the OverFeat [21] method of the AlexNet structure and the current advanced ReInspect [8] method. ReInspect has 3 variants, using different types of loss functions, named L_fix and L_firstx and L_hungarian. Our will be compared with all the methods mentioned above to evaluate its performance. The following Table 1 gives a summary of test result.

As shown in Table 1, our model has achieved an AP result of 0.69, which is higher than other three types of benchmark methods, and close to the performance of the best method (ReInspect-L_hungarian).

Table 1. Performance on brainwash dataset

Dataset	Scenario	Method	AP
Brainwash	Crowded scenes (7.89 avg. head count)	Overfeat-AlexNet	0.61
		ReInspect, L_fix	0.59
		ReInspect, L_firsts	0.62
		ReInspect, L_hungarian	0.77
		Ours	**0.69**

4.2 Result on Hollywoodheads Dataset

To evaluate the performance of our model on the HollywoodHeads dataset, a new model was retrained using the training set of the dataset. After training, test the performance of the model on a test set containing 1297 images. At the same time, our model is compared with the methods mentioned in [7], including the head detector method based on DPM [22], the head detector method based on R-CNN [23], and the methods in [22] including some variant versions named the local context model method (Context-Local), the local + global + paired combined context model method (Context-L + G + P), The following Table 2 gives a summary of test result.

Table 2. Performance on Hollywoodheads dataset

Dataset	Scenario	Method	AP
Hollywoodheads	Movie scenes (1.65 avg. head count)	DPM Face	0.37
		R-CNN	0.67
		Context – Local	0.71
		Context Local + Global + Pairwise	0.72
		Ours	**0.73**

Compared with all benchmark methods, our method has achieved the best head detection results. Compared with the method proposed in [22] and its variant methods, it is still better than the best performing method in AP index about 1%.

4.3 Ablation Experiment

In order to prove that the anchor size set to our model is useful, we also carried out an ablation experiment. In this experiment, three models with different anchor size were trained on Brainwash dataset, including (1) $32 \times 32, 64 \times 64$ (Default setting); (2) $64 \times 64, 128 \times 128$; (3) $128 \times 128, 256 \times 256$. These models are all trained with the same hyper parameters. The following Table 3 gives a summary of test result.

Table 3. AP results with different anchor size

Anchor size	AP
$32 \times 32, 64 \times 64$	0.69
$64 \times 64, 128 \times 128$	0.53
$128 \times 128, 256 \times 256$	0.45

It can be seen from Table 3 that when the anchor size is 32×32 and 64×64, the best AP results will be obtained, which proves the conclusion we proposed in the previous sections that choosing the anchor size according to the effective receiving domain can get better detection results.

5 Conclusion

In this work, we proposed a fully convolutional end-to-end trainable head detection model based on anchors, which is suitable for head detection in crowded indoor scenes. Our model is tested in the challenging head detection dataset (Brainwash) and achieve an AP of 0.69, which is comparable to some previous benchmark methods. In the future, we plan to improve our work by increasing the overall accuracy, especially in the case of smaller head size, which may be done by combining from detection of various convolutional layers.

Acknowledgments. This work was supported by the National Educational Science 13th Five-Year Plan Project (JYKYB2019012) and the Basic Research Fund of the Engineering University of PAP (WJY201907).

References

1. Zhang, C., Li, H., Wang, X.: Cross-scene crowd counting via deep convolutional neural networks. In: Proceedings of the IEEE Conference on Computer Vision and Pattern Recognition, pp. 833–841 (2015)
2. Zhang, Y., Zhou, D., Chen, S.: Single-image crowd counting via multi-column convolutional neural network. In: Proceedings of the IEEE Conference on Computer Vision and Pattern Recognition, pp. 589–597 (2016)
3. Sam, D.B., Surya, S., Babu, R.V.: Switching convolutional neural network for crowd counting. In: Proceedings of the IEEE Conference on Computer Vision and Pattern Recognition, pp. 4031–4039 (2017)
4. Mae, Y., Sasao, N., Sakaguchi, Y.: Head detection and tracking for monitoring human behaviors. In: Systems and Human Science Conference, pp. 477–487 (2005)
5. Zhang, D., Yafei, L., Liwei, H., Peng, H.: Multi-human tracking in crowds based on head detection and energy optimization. Inf. Technol. J. **12**(8), 1579–1585 (2013). https://doi.org/10.3923/itj.2013.1579.1585

6. Merad, D., Aziz, K.-E., Thome, N.: Fast people counting using head detection from skeleton graph. In: 7th IEEE International Conference on Advanced Video and Signal Based Surveillance, pp. 233–240 (2010)

7. Vu, T.-H., Osokin, A., Laptev, I.: Context-aware cnns for person head detection. In: Proceedings of the IEEE International Conference on Computer Vision, pp. 2893–2901 (2015)

8. Stewart, R., Andriluka, M., Ng, A.Y.: End-to-end people detection in crowded scenes. In: Proceedings of the IEEE Conference on Computer Vision and Pattern Recognition, pp. 2325–2333 (2016)

9. Marin-Jimenez, M.J., Zisserman, A., Eichner, M., Ferrari, V.: Detecting people looking at each other in videos. Int. J. Comput. Vision **106**(3), 282–296 (2013). https://doi.org/10.1007/s11263-013-0655-7

10. Subburaman, V.B., Descamps, A., Carincotte, C.: Counting people in the crowd using a generic head detector. In: IEEE Ninth Inter-national Conference on Advanced Video and Signal-Based Surveillance, pp. 470–475 (2012)

11. Felzenszwalb, P.F., Girshick, R.B., McAllester, D.: Object detection with discriminatively trained part-based models. IEEE Trans. Pattern Anal. Mach. Intell. J. **32**(9), 1627–1645 (2010)

12. Ren, S., He, K., Girshick, R.: Faster R-CNN: towards real-time object detection with region proposal networks. IEEE Trans. Pattern Anal. Mach. Intell. J. **1**(6), 1137–1149 (2017)

13. Krizhevsky, A., Sutskever, I., Hinton, G.E.: Imagenet classification with deep convolutional neural networks. Adv. Neural Inf. Process. Syst. Conf. **25**, 1097–1105 (2012)

14. Szegedy, C., Liu, W., Jia, Y.: Going deeper with convolutions. In: Proceedings of the IEEE Conference on Computer Vision and Pattern Recognition, pp. 1–9 (2015)

15. https://arxiv.org/abs/1409.1556. Accessed 4 Sep 2014

16. Luo, W., Li, Y., Urtasun, R.: Understanding the effective receptive field in deep convolutional neural networks. In: Advances in Neural Information Processing Systems Conference, pp. 4898–4906 (2016)

17. Girshick, R.: Fast R-CNN. In: Proceedings of the IEEE International Conference on Computer Vision, pp. 1440–1448 (2015)

18. Russakovsky, O., et al.: Imagenet large scale visual recognition challenge. Int. J. Comput. Vision **115**(3), 211–252 (2015). https://doi.org/10.1007/s11263-015-0816-y

19. Paszke, A., Gross, S., Chintala, S.: Automatic differentiation in pytorch. In: NIPS Conference, pp. 1–4 (2017)

20. Everingham, M., Eslami, S.A., Van Gool, L.: The pascal visual object classes challenge: a retrospective. Int. J. Comput. Vision **111**(1), 98–136 (2015)

21. https://arxiv.org/abs/1312.6229

22. Mathias, M., Benenson, R., Pedersoli, M., Van Gool, L.: Face detection without bells and whistles. In: Fleet, D., Pajdla, T., Schiele, B., Tuytelaars, T. (eds.) ECCV 2014. LNCS, vol. 8692, pp. 720–735. Springer, Cham (2014). https://doi.org/10.1007/978-3-319-10593-2_47

23. Girshick, R., Donahue, J., Darrell, T.: Rich feature hierarchies for accurate object detection and semantic segmentation. In: Proceedings of IEEE Conference on CVPR, pp. 580–587 (2014)

Network Communication and Security

A Survey on the Classic Active Measurement Methods for IPv6

Yipeng Wang, Tongqing Zhou$^{(\boxtimes)}$, Changsheng Hou, Bingnan Hou,
and Zhiping Cai

College of Computer, National University of Defense Technology, Changsha, China
{wangyipeng,zhoutongqing,zpcai}@nudt.edu.cn

Abstract. With the rapid development of the Internet, the IP network
has become the infrastructure of human society and plays an important
role in life. However, the IPv4 network is criticized as it faces address
exhaustion problem. At the same time, there are many problems in the
IPv4 network in the process of practical use, such as routing table expansion, fragile security, etc.

All these have prompted the network infrastructure to upgrade to
the IPv6 network. In order to measure and monitor the IPv6 network,
new methods and techniques are required as those tailored to the IPv4
network cannot be adaptively applied here. Active measurement can
obtain network path distribution and traffic-related information, thus is
commonly used in network routing behavior analysis, network topology
detection and visualization. To this end, the design of appropriate and
efficient active measurement method for the IPv6 network is an essential
issue that has been widely studied.

In this paper, we present a survey on the active measurement of
IPv6 network. A technique-based taxonomy is presented by introducing
and analyzing the tunnel discovery, address detection, network topology
discovery, and bandwidth measurement methods for the IPv6 network.
The paper also discussed the unsolved problems, challenges and future
research directions of IPv6 active measurement.

Keywords: IPv6 · Active measurement · Tunnel and topology
discovery · Address detection · Bandwidth measurement

1 Introduction

The Internet has become an essential infrastructure in our daily lives. As the
next-generation IP protocol, the IPv6 is designed by the Internet Engineering
Task Force (IETF) to replace IPv4. In early 1992, IETF put forward the idea
of a new generation of Internet address system. In 2003, IETF released the first
IPv6 network (i.e., 6bone network) used by the IETF IPng (IP Next Generation)
working group to test the IPv6 networks [27]. It mainly tests how to migrate the
IPv4 network to the IPv6 network. As a platform for problem testing, the 6bone
network includes realizing the IPv6 network protocol, the migration of IPv4 to

© Springer Nature Singapore Pte Ltd. 2021
K. He et al. (Eds.): NCTCS 2020, CCIS 1352, pp. 149–171, 2021.
https://doi.org/10.1007/978-981-16-1877-2_11

IPv6 and other functions. Basically, IPv6 network shows seven aspects of advantages compared with the IPv4 network: it has a larger address space, smaller routing table, better header format, enhanced multicast support, Flow Control [1], auto configuration support, and expansion of the protocol. As an ungraded version, the IPv6 network is not forward compatible with IPv4 in nature, while is compatible with some auxiliary Internet protocols, such as Transmission Control Protocol (TCP), User Datagram Protocol (UDP), Internet Control Message Protocol (ICMP), Internet Group Management Protocol (IGMP), Open Shortest Path First (OSPF), Border Gateway Protocol (BGP) and Domain Name System (DNS) [2]. Development of IPv6 has gone through 20 years, along with the investigation on effective measuring the IPv6 network indicators.

With the acceleration of the IPv6 network construction, two fundamental problems emerge: 1) How to provide experimental and technical basis for network technology research so that the next-generation network can better meet the country's needs has become a very important task [8]? 2) How to scientifically evaluate the stability and reliability of the network requires correct measurement of data such as delay, packet loss rate, and bandwidth reflecting network operation and performance?

As a traditional scanning method, active measurement technique obtains network performance and behavior parameters by injecting specific measurement messages into the network and analyzes the impact on the target network and the resulting characteristic changes. With the injection of traffic, we are expected to get impressive responses and certain performance feedbacks from the network. In this way, end-to-end performance parameters between selected network endpoints can be obtained in a convenient and flexible manner [3]. Meanwhile, active measurement strategy is known to be user-friendly since it does not involve user traffic information that may affect user privacy [4]. However, active measurement needs to inject additional measurement traffic into the network, which increases the network load and affects user traffic [49]. From this point of view, one should take balance between measurement utility and traffic burden when performing active injection. Over the years, there have been several calls from within the Internet measurement community to summarize the critical core methods of network measurement to promote technological progress. Our study finds inspiration in these prior works.

Although various application directions and methods for active measurement can be found in literature, they are all aimed at a certain direction and there is no comprehensive summary of active measurement technology. Active measurement in IPv6 networks is the first step towards understanding the network model and moving towards the future. In this paper, we present a comprehensive overview of the active measurement research for IPv6 network. A technique-based taxonomy is described by introducing and analyzing the tunnel discovery, address detection, network topology discovery, and bandwidth measurement methods for the IPv6 network. We point out that this four aspects of measurement task are essential for understanding the operating status and efficiently organizing IPv6 network. Meanwhile, we also provide open issues, research challenges, future research directions, and potential areas for future research to promote IPv6 active measurement.

The rest of this paper is consequently organized as follows. Section 2 introduces IPv6 tunnel method. Section 3 introduces IPv6 address detection. Section 4 introduces IPv6 network topology discovery. Section 5 introduces IPv6 bandwidth measurement. Open issues, research challenges, and future directions are addressed in Sect. 6. Finally, the paper is concluded in Sect. 7.

2   IPv6 Tunnel Technology

Tunneling technology is used in the transition from IPv4 network to IPv6 network and is a standard established for network convergence. It is essentially a multi-layer superposition of the message header. Although it can enable the IPv4 network to be interconnected with the IPv6 network, this dramatically reduces the network forwarding efficiency because the message content must include two message headers, regardless of whether it is IPv6-Over-IPv4 or IPv4-Over-IPv6 [50].

Discovery on the available tunnels is essential for establishing communication, thus making it an important measurement task in the IPv6 network. In this section, we first introduce the typical modes of IPv6 tunnel, then describe the classic tunnel discovery technique.

2.1   Typical Modes for IPv6 Tunnel

The tunnel technology carries internal data packets across a specific network by encapsulating outer packet headers. The commonly used encapsulation forms of tunnels are mainly categorized into two types: 1) IPv6 data packets are directly encapsulated into IPv4 packets, such as 6to4 and ISATAP; 2) IPv6 packets are encapsulated into IPv4+UDP, such as Teredo and TSP. The main difference between the two modes is that the method in which IPv6 packets are directly encapsulated into IPv4 packets is generally used in public network environments. In contrast, the method in which IPv6 packets are encapsulated in IPv4+UDP can be used in complex environments such as NAT. The following two classic tunnel technologies, 6to4 and Teredo, are taken as examples to study the above two modes.

Directly Encapsulated. This mode treats IPv6 data packets as a part of IPv4 data packets, as shown in Fig. 1. Wherein, "Tunneled Packet" means the inner packet and "Tunneling Packet" means the outer packet. Since IPv4 data packets directly encapsulate IPv6 data packets, they can be easily identified and filtered by routers. However, this mode is not suitable for complex network environments such as NAT [9], since the required translation information is not filled in the header.

The 6to4 tunnel is a typical tunnel of this model, including 4 entities: 6to4Host, 6to4Router, 6to4Host/Router and 6to4Relay. The 6to4Host takes IPv6 as a terminal and is located inside a network managed by a 6to4Router, which is inside the system operated by the 6to4Router and generally does not have a public network IPv4 address. The 6to4Router is the boundary device of the 6to4

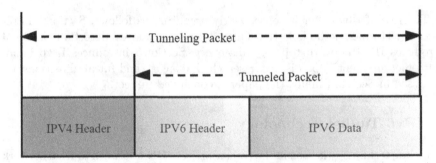

Fig. 1. IPv6 tunnel of directly encapsulated mode [50]

tunnel, which is on the edge of the IPv4 network and has a public network IPv4 address. It automatically configures IPv6 connections for 6to4Hosts inside the network, encapsulates and decapsulates data packets and realizes the communication with other 6to4Routers or 6to4Relays. The 6to4Host/Router combines the features of the 6to4Host and the 6to4Router and requires a public network IPv4 address. The 6to4Relay is the connection point between IPv4 Internet and IPv6 Internet. An example scenario of the 6to4 tunnel is shown in Fig. 2, where two IPv6 hosts communicates on the IPv4 network using thetunnelling devices.

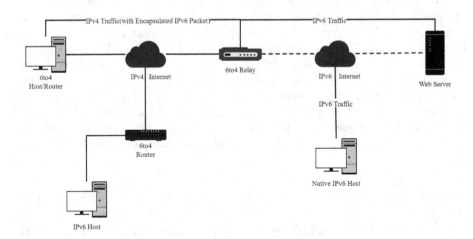

Fig. 2. Typical topology of 6to4 tunnel

Encapsulated into IPv4+UDP. In this mode, IPv6 packets are encapsulated in the form of UDP+IPv4 headers. Due to the use of UDP protocol for encapsulation, it is not easy to be detected by routers and can be used in complex network environments such as NAT. The packaging mode is shown in Fig. 3.

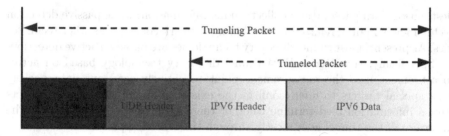

Fig. 3. IPv6 tunnel of encapsulated into IPv4+UDP mode [50]

Teredo tunnel is one of the typical tunnels of this encapsulation mode, which contains 3 primary devices: TeredoHost, TeredoServer and TeredoRelay. TeredoHost is an IPv6/IPv4 dual-stack terminal user that supports the Teredo tunnel interface, which is generally set up inside the NAT subnet. TeredoServer provides automatic address configuration for TeredoHosts, and at the same time offers initial routes for TeredoHosts to other TeredoHosts or IPv6 networks. TeredoRelay provides real-time data forwarding service between TeredoHost and the IPv6 network. Both TeredoServers and TeredoRelays are located at the IPv4 and IPv6 Internet connection points to encapsulate and decapsulate data packets. The Teredo network topology and device deployment location are shown in Fig. 4.

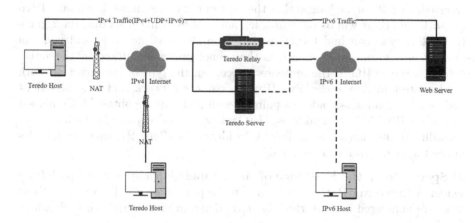

Fig. 4. Typical topology of Teredo tunnel

2.2   IPv6 Tunnel Discovery Technology

The IPv6 tunnel discovery technology can generally be divided into active detection technology and passive detection technology. Active detection technology is to speculate and confirm the existence of the tunnel by sending special destination data packets, and collect tunnel information; while passive detection technology is mainly realized by combining network control and other technologies. At present, most tunnel discovery technologies are based on active detection.

Specifically, the IPv6-in-IPv4 tunnel discovery technology based on active measurement uses the measurement point to actively send out data packets with special targets to infer, confirm the existence of the tunnel, and collects tunnel information to determine whether there is an IPv6-in-IPv4 tunnel. The discovery of the tunnel goes mostly through the following processes:

1) Looking for Dual-Stack Nodes. Dual-stack node is a necessary condition for the existence of IPv6 tunnel. The IPv6 network provides a multicast address FF02::1. For datagrams with a destination address of FF02::1, all nodes (including routers, hosts, etc.) in the link-local range will receive and process them; besides, it also provides an FF02::2 address, for datagrams with the destination address FF02::2, all routers in the link-local range will receive and process them, but the host will ignore it. Therefore, searching for dual-stack nodes in the target network can be achieved through the following two steps:

The first step is to send a data message to FF02::1. If there is no response, it can be determined that the type of the target network is a pure IPv4 network and there is no dual-stack node; if there is a response, then it can be determined that the type of the target network is not IPv4. And the <IPv6 link-local address, MAC address> address pairs of each node can be obtained, then proceed to the second step.

The second step is to use ARP to poll all nodes in the subnet. If no response is received, then it can be judged that the type of the target network is a pure IPv6 network, and there will be no dual-stack nodes. After further sending datagrams to FF02::2, you can find the IPv6 router on the IPv6 network, and you can also get critical information such as the subnet prefix announced by the active detection router [10]. If the response is received, the type of the target network can be determined to be the IPv6/IPv4 coexistence network, and then the <IPv4 address, MAC address> address pairs of each node can be obtained. Combined with the <IPv6 link-local address, MAC address> address pair in the first step, according to the uniqueness of the MAC address, the IPv6/IPv4 dual-stack nodes in the target network can be found.

2) Speculate on the Existence of the Tunnel. Suppose there is a path from sender A to receiver B, then each link on the path has an MTU value. If these links are numbered in turn, the schematic diagram of the first link is shown in Fig. 5.

Fig. 5. The i-th link on the path from node A to node B

Assuming that the interface at both ends of the i-th link is X(i) and Y(i), and the MTU value on the link is MTU(i), the path MTU discovery technology can be used to obtain the MTU of each path from A to i are denoted as PMTU(i). According to the definition of PMTU, we can have:

$$PMTU(i) = min(MTU(i), PMTU(i-1)) \qquad (1)$$

Since IPv6 tunnel needs to encapsulate IPv6 packets, the MTU value on the tunnel will be smaller than that of the lower-layer IPv4 network, which makes different types of IPv6 tunnels have different MTU values. For IPv6 tunnels that adopted directly encapsulation, the overhead of the encapsulation header (IPv4 header) is 20 bytes; for GRE tunnels, the overhead of the encapsulation header is 24 or 28 bytes. So, we can use such a PMTU variation relationship to speculate whether there may be a tunnel on a path, namely, if $PMTU(i) < PMTU(i-1)$, and the value of PMTU(i) is a few special values, there may be a tunnel (X (i), Y (i)) between port X(i) and Y(i).

3) Confirm the Existence of the Tunnel. Suppose it has been speculated that there may be a tunnel between nodes A and B, and the IPv4 addresses of A and B are known to be A4 and B4, as shown in Fig. 6. Then we can actively send data packets to confirm the existence of the tunnel.

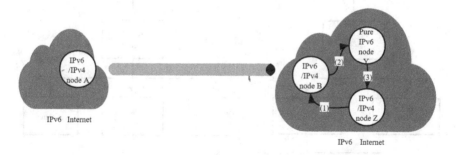

Fig. 6. Schematic diagram confirming the existence of the tunnel

First, an IPv6 tunnel datagram can be sent from an IPv6/IPv4 dual-stack node Z to a known pure IPv6 node Y (assuming its address is Y6) in the IPv6 network through the IPv4 network. After the datagram arrives at node B, if there is no tunnel between nodes A and B, then node B will directly discard the datagram; if there is an IPv6 configuration tunnel between nodes A and B, the datagram can pass the inspection of node B and be disassembled into IPv6 Datagram by node B. Next, the IPv6 datagram is sent from node B to the destination address Y6 via the IPv6 network. Finally, after node Y receives the datagram, it sends an IPv6 response datagram, which is directly transmitted to node Z via the IPv6 network. According to whether the third-party IPv6/IPv4 node Z has received the ICMPv6 response message from the pure IPv6 node Y, it can be determined whether there is an IPv6 configuration tunnel between nodes A and B.

3   IPv6 Address Detection

Address detection is a technology that collects, counts, and analyzes network information based on certain basic knowledge of network address allocation and big data detection algorithms. This technology is of great significance to network measurement. Through this technology, network information can be collected more accurately and network vulnerabilities can be found.

The classic IPv6 address detection methods (non-violent scanning) is classified into active detection and passive detection as shown in Fig. 7. For active measurement, it contains the technique based on DNS (Domain Name System) data and the method based on seed address. The most crucial difference between them is whether the collected address will be further analyzed, and a new address will be generated for active detection [23].

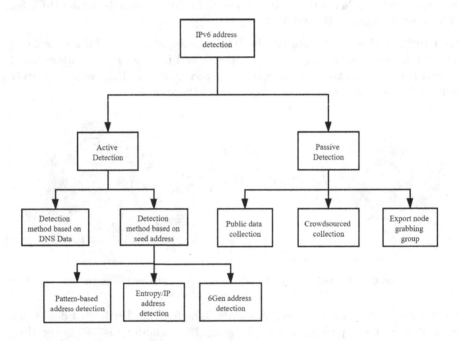

Fig. 7. Classification of IPv6 address detection technologies [23]

In this section, we first briefly introduce the typical active detection methods, then describe the seed address based address detection category in details.

3.1   Overview on Active Address Detection

Collecting IPv6 addresses through DNS is an effective method [11]. The form based on DNS data mainly collects IPv6 addresses by actively sending DNS query data packets [12]. Strowes [6] proposed a method of reverse-checking the

IPv4 address and then performing AAAA query to get the IPv6 address. Subsequently, Fiebig [13] et al. collected $2.8*10^6$ IPv6 addresses through further use of DNS reverse lookup (PTR records). The advantage of collecting IPv6 addresses through DNS is that many of them are addresses of devices that provide services. Still, the disadvantage is that the number of IPv6 addresses discovered is quite limited. Besides, Borgolte [14] et al. proposed a method of collecting IPv6 addresses in combination with the security extension of DNSSEC (Domain Name System Security Extensions) system. Still, in many cases, there are not enough conditions for passive collection. In view of these problems, the detection method based on the seed address is introduced.

Seed addresses are IPv6 addresses that have been or have been surviving collected by researchers. By mining the inherent law of existing IPv6 address allocation [16], this type of detection technology selectively generates new IPv6 addresses as the input of active scanning. Generally, it needs to rely on other methods to provide IPv6 seed addresses as input to the algorithm, which can be regarded as an extension of different ways. The detection method based on the seed address can detect a large number of surviving IPv6 addresses in the vast IPv6 address space, which is a very efficient detection method [17].

Seed-based address detection technology requires an existing IPv6 seed address as input, and generates new IPv6 addresses that may survive through modelling and analysis, thereby reducing the search space, and then verifying these new IPv6 addresses through active detection whether to survive. There are mainly the following classic address detection methods, and other methods are derived from three methods.

3.2   Pattern-Based Address Detection

This method can be understood as finding the rules in the IPv6 address set; for example, the IPv6 interface identification allocation method can be counted as a pattern. Early IPv6 detection work is mainly to generate new addresses by analyzing the rules that are easy to guess in IPv6, which can be combined with some artificially set rules to achieve IPv6 address detection. These rules include Low-Bytes, EUI-64 embedding, IPv4 embedding and so on [18], also proposed a lot of standard IPv6 address generation methods. The idea of early IPv6 address detection is to mine the typical characteristics of the 128bit IPv6 address itself and guess the IPv6 address that is most likely to be used or allocated.

Many early types of research on the mining of IPv6 address rules were based on experience to make artificial guesses and artificially set new regulations to discover new IPv6 addresses. But the Pattern-based address detection technology takes a different approach, which can automatically analyze the patterns of the seed address set and generate new IPv6 addresses based on these patterns [17]. The core of Pattern-based address detection technology is a recursive address generation algorithm. It starts with the bit set initially, and then every time a new bit is added recursively to the pattern, the program stops when the bit that is not included in the way reaches the set threshold [19]. The whole algorithm mainly contains four links as follows:

– Extract pattern: The final extracted design is the one with the most matching
  seed addresses among all candidate patterns. In each recursive search process,
  the value (0 or 1) in the pattern is determined according to the value of the
  seed address in this bit.
– Reverse rule: the full recursive search is similar to a binary search tree.
– Initialization: initialization needs to specify X bits; X is a positive integer,
  and the value is at least 1. The algorithm starts the search with a value of
  2X to ensure that all possible search spaces are covered.
– Termination condition: if the number of bits that have not been added to the
  pattern reaches the set threshold, the algorithm stops.

This research work created a precedent for heuristic IPv6 address detection
technology. Pattern-based address detection technology has two main advantages: First, it can automatically mine the value of each bit of the address from
the seed address, avoiding the defects of human analysis, and can successfully
apply its method to new seed addresses in the collection. Second, the number of
addresses discovered by this technology is higher than that of violent scanning,
confirming the conclusion that the heuristic detection method for seed addresses
is better than violent scanning.

But Pattern-based address detection technology has strict requirements for
seed address sets. If the randomization of the seed address is severe, the hit rate
will be significantly reduced, which will seriously affect the effect of Pattern-
based address detection technology. Also, its most significant contribution is to
cause researchers to start thinking about using heuristic methods to analyze seed
addresses, looking for rules from known addresses to detect the unknown and
huge IPv6 address space.

3.3   Entropy/IP Address Detection

This is a method of automatically mining the seed address structure [20]. This
technology combines Shannon's entropy of information theory, machine learning
and probabilistic graph models, and provides a solution to visualize the IPv6
address structure. Data visualization can help researchers to understand the
characteristics of IPv6 address sets as a whole. The address generation algorithm
of Entropy/IP address detection technology can be divided into three steps:

1) Shannon Entropy. Shannon entropy is used to solve the problem of quan-
titative measurement of information [21]. The calculation principle of Shannon
entropy is shown in formula (2). The value range of Shannon entropy is [0,1].
The higher the entropy value, the greater the uncertainty of the variable and the
more potent its randomness.

$$H\left(x\right) = -\sum_{i=1}^{k} P\left(x_i\right) lb_P\left(x_i\right) \tag{2}$$

Entropy/IP-based address detection technology calculates the entropy value of
each nibble in the IPv6 set.

2) Address Segmentation. An IPv6 address has 32 nibbles (128 bits) in total, this step combines the nibbles with similar entropy values into segments. The equivalent threshold cannot be set too large or too small; otherwise, it will seriously affect the correctness of the segmentation, so that the subsequent Bayesian modelling effect is low.

3) Bayesian Network Modelling. The Bayesian graph model of the complete address is constructed according to the conditional probability between segments. This graph model describes the value of each segment and the conditional probability of the value of each segment and the previous segment. Finally, it is possible to randomly traverse the entire value space described by the Bayesian probability model from the Bayesian network diagram, and select the number of addresses set by the user as the target to be detected.

Entropy/IP technology applies Bayesian and information theory knowledge to the field of IPv6 address detection for the first time, and proposes a heuristic address production model that provides an effective method for visualizing IPv6 address structure. This technology also has a useful application in continuous detection scenarios, because once Entropy/IP completes Bayesian modelling, the next address generation part does not need to be remodelled.

Although this technology solves the IPv6 address structure visualization problem well and also has the characteristics of exploring the seed address from a global perspective, the address hit rate generated by the algorithm is low. With the randomization of the current address IID allocation method, this algorithm is likely to have an entropy value close to 1 for the nibble calculation of the 64-bit part of the IPv6 address, so the latter 64 bits may be modelled by Bayesian as a whole segment or several segments, seriously affecting the effectiveness of the algorithm.

3.4   6Gen Address Detection

6Gen identifies the dense areas in the address value space from the seed address set and select new address values in these dense areas to generate candidate addresses that need to be scanned. The current seed address is also the most likely value area of unknown address in the region with the densest values of different nibbles. The core steps of the address generation algorithm for 6Gen address detection technology are shown in Fig. 8.

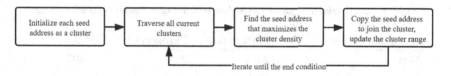

Fig. 8. The core steps of the 6Gen algorithm [23]

6Gen uses the DBSCAN (Density-Based Spatial Clustering of Applications with Noise) clustering algorithm for the set of input IPv6 seed addresses, which initially treats each seed address as a cluster. In the iterative process, the range of multiple cluster representatives is continuously increased until the threshold is reached. 6Gen's address generation algorithm has two crucial concepts-clustering range and density.

- Clustering range [22]: The clustering range is determined by all the seed addresses of the cluster set at different nibble positions. There are two modes of range definition: Tight and Loose. The difference lies in whether the nibbles that deal with changes use the maximum and minimum values as the range or [0-f] all possible values as the range.
- Clustering density: The clustering density is equal to the ratio of the number of seed addresses in the clustering set to the clustering range.

6Gen address detection technology is a relatively successful technology that applies the clustering algorithm to the IPv6 detection field. This technology studies the potential address values in the IPv6 address space by characterizing the relationship between the IPv6 seed address sets. The 6Gen address detection technology first clusters existing IPv6 addresses to find the "area" where these addresses are located and then searches for undiscovered IPv6 addresses in these areas [24].

6Gen can discover more IPv6 addresses, and the hit rate of IPv6 address detection generated by 6Gen on a smaller address set is higher than that of Entropy/IP address generation technology. Another contribution of 6Gen is to point out the problem of alias addresses and alias prefixes of IPv6 generated addresses for the first time [25]. The existence of a large number of alias addresses in the seed set will seriously affect the universality of the address generation algorithm and will make the addresses generated by the algorithm "biased", that is, most of the surviving addresses detected are likely to point to the same interface. For this reason, 6Gen proposes a simple random detection method to find alias prefixes [6]. The design idea is straightforward and intuitive. Since a large number of IPv6 alias addresses point to the same interface, a certain number of addresses are randomly generated under the alias prefix. If it is detected as alive, then these addresses are likely to be alias addresses, and this prefix is the alias prefix. Because in 6Gen's view, the IPv6 address space under a 64-bit prefix is so huge, and IPv6 addresses are so sparse, it is impossible for these randomly

Table 1. Comparison of three detection technologies based on seed address

Detection method	Pattern-based	Entropy/IP	6Gen
Year	2015	2016	2017
Time consuming	Fastest	Medium	Slowest
Core technology	Recursion	Bayesian network	Clustering
Whether to consider aliases	No	No	Yes
Visualization	Weaker	Stronger	Weaker

generated addresses to survive. However, the 6Gen algorithm itself is continuously increasing the set of clusters, and there is a large amount of intersection between each cluster set, which makes the memory consumption of the whole algorithm relatively large. With the increase in the number of addresses, the time consumed to detect addresses will increase exponentially [5].

4 IPv6 Network Topology Discovery

The active measurement method of network topology discovery is to randomly send small detection messages to the network through the detection end, examine the delay characteristics of the messages in the network, use the link-state statistical algorithm to get them active and idle status, and finally get the link available bandwidth [26]. It can also analyze the traffic changes of each routing node. Compared with the traditional packet pair method, this method is not limited by the maximum sending rate of the source node and has lower requirements for data post-processing and filtering [15].

Network topology measurement should have the following characteristics: (1) High efficiency, network topology discovery should be to obtain network topology information in the shortest time and with the smallest network load; (2) Integrity, network topology measurement should try its best to obtain the network The relationship between nodes; (3) Accuracy, network measurement and generated topology should be consistent with the real network topology.

4.1 Levels of Network Topology Discovery

According to granularity, the network topology can be divided into autonomous system (AS) level topology, IP (or router) level topology. Different topology types have corresponding topology discovery technologies [28].

AS-level Topology Discovery. The AS-level topology is a network topology structure composed of autonomous system nodes and edges represented by BGP connections [29]. In the AS-level topology diagram, a node represents an AS, and the connection between nodes does not refer to the actual relationship between routers but refers to whether there is a BGP session between AS. Although it is only a coarse-grained topology, it is an essential means to study the Internet model.

The classic techniques of AS topology discovery mainly include BGP routing information analysis method and Traceroute detection method. The method based on BGP routing information analysis constructs the AS-level topology map through the AS-Path attribute information in the BGP routing table or the BGP update message. The method based on Traceroute detection first uses Traceroute detection to obtain the IP-level forwarding path, then maps the IP address to the AS to which it belongs to obtain the AS-level forwarding path, and finally merges the same AS nodes to obtain the AS-level topology.

Router-Level Topology Discovery. Router-level topology is divided into backbone network router-level network topology and access network layer topology according to different research fields. The backbone router-level topology is mainly a topology composed of routers in the backbone network. It can analyze the operating status of the IPv6 network protocol and is of great significance for studying the characteristics and evolution of the actual network. The access network layer topology is mainly used for the discovery and management of the access network layer devices (mainly routers) and the management of the IP network. Specifically, it is necessary to discover the connection relationship between routers, subnets, routers and routers, and routers and subnets [30].

Since the purpose of discovering the network layer of the backbone network and the access network is different, the corresponding discovery methods are also different. Backbone network topology discovery has a wide detection range but requires low accuracy. It usually uses general protocols (UDP, ICMP, etc.) to obtain topology information; access network topology discovery has a small detection range, but requires high accuracy, and requires unique network management protocol (SNMP protocol, etc.) to obtain critical configuration information of network devices (such as the number of interfaces, device names, routing information, etc.).

The principle of the network topology discovery algorithm based on the general protocol is to first send an ICMP echo request to an IP address in a specified address space, determine the reachability of the devices in the network, and discover active nodes; then use Traceroute and depth-first traversal strategies to detect the discovery. It uses ICMP timeout and unreachable port information to find the routing information from the source node to the target node; finally, analyze the obtained data to construct the network topology connection. The basic idea of the SNMP-based network-layer topology discovery algorithm is to start from the default router of the subnet where the network management workstation is located, and obtain the information related to the network topology in the MIB through the SNMP protocol, such as the subnet information and the next hop in the routing table Information (IP Route Next Hop [31]); Then, the breadth-first traversal method is adopted to traverse the routers in the network, and the routers and subnets in the network are gradually discovered downward. Finally, the information is analyzed and processed to obtain the topology structure.

4.2   Hierarchical Topology Discovery Method

This algorithm separates the topology discovery of the backbone network from the topology discovery in the subnets. It distributes them on different measurement points to form a topology discovery structure of "management centre-local agent" mode. The topology discovery program of the backbone network is arranged on the management centre, which is responsible for discovering the routers in the backbone network and the connections between them. Several detection points are placed in the network (one detection point is arranged in each subnet), each detection point has a local agent, and the topology discovery program of the subnet is set on the local agent, responsible for discovering the nodes and configuration in the subnet information. The local agent reports the topology information in the subnet and the connection relationship between the local agent and the management centre to the management centre.

The specific principles and main functional modules of this method are as follows:

- Local agent. The task of the local agent is to discover the local link address, MAC address, node type (host or router), subnet prefix and the IPv6 address hostname available to all nodes. An LA only manages one subnet; for a multi-interface device, the function of each LA can only be bound to one interface. A device can support multiple LA [32].

  The operation of LA adopts two modes: active and passive [33]. Actively obtain the local link address related to the MAC address by sending ICMPv6 and neighbour solicitation messages. The disadvantages are: there are conflict domains and broadcast storm problems; it is impossible to obtain a global IPv6 address. The passive method realizes the acquisition of link information by monitoring the transmission of data on the subnet and intercepting neighbour advertisement messages. The destination address of the neighbour advertisement message is at least the global IPv6 address of a host in the subnet. However, neighbour broadcast messages are sent regularly, and the information of the local agent may not reflect the current information in the subnet [35].

- Central manager. The task of the central manager is to collect the information sent from the local agent, find the backbone network, and aggregate the information to update the topology. The address of the management centre is an IPv6 anycast address, and data packets sent to this address will only be received by the first node (implemented by a specific routing strategy).

- Communication protocol. The information exchange between the local agent and the management centre can be through a simple asynchronous UDP protocol, or the trap mechanism(Trap Mechanism) of SNMP [34]. Using this topology discovery structure can separate the discovery of the subnet part from the entire system and distribute it on multiple detection points. By using this topology discovery structure, the discovery of the subnet can be separated from the whole system and distributed in various detection points, each of point sends a group of multicast addresses to discover the address information and host information of all nodes in the subnet. In this way, the node information in multiple subnets can be obtained, which makes the results of topology discovery more comprehensive and detailed. The Management Centre (CM) centrally processes the subnet information sent by multiple local agents and uses Traceroute to realize the structure discovery of the backbone network, and then analyzes the topology of the entire network.

However, at present, there are still some difficulties in active measurement of large-scale network topology, which can be summarized as follows [36]:

- Internet network is composed of many networks with different structures. Different network devices and software are running in various networks. And

for the consideration of security policy, various networks will set some filtering rules, which bring difficulties to network topology measurement;
- Although most of the network node devices in the network support essential services, these network devices are suitable for a foolproof topology discovery method. However, some network devices in the network still do not follow these due to security considerations. Resulting in insufficient integrity of network topology discovery;
- Dynamic changes in the network. Now for a large-scale network, it may take a long time to perform a topology measurement, and the detected topology may be outdated, so the detection is meaningless;
- There is no real network topology data for verification. For security reasons, operators are generally reluctant to disclose their network topology. At the same time, network topology detection is also restricted by technology, law and social factors, making the network topology detection technology not mature enough.

5   IPv6 Bandwidth Measurement

Network bandwidth is a vital reference indicator for measuring network connection capabilities, and it is also the primary pricing basis for ISP (Internet Service Provider) for Internet access charges. Because higher bandwidth usually means higher data throughput and better service quality [37]. In this case, the appropriate measurement of network bandwidth has become a critical task. For users, they need to know whether they have fully utilized their access bandwidth. For ISP, they need to use network bandwidth measurement tools to understand the real-time operating status of the network, and to find out the network congestion points and underutilized links in time. For Internet applications, such as content distribution system, streaming media system and multi-person video conference system, we can change the bit flow rate adaptively by measuring the end-to-end bandwidth to improve the program performance [38]. More importantly, it can guide network protocol developers, adopt new algorithms to control routing choices and formulate better congestion control strategies. Network bandwidth can generally be divided into end-to-end path capacity, end-to-end available bandwidth and nominal bandwidth. The basis of the network end-to-end path capacity measurement method is single-packet technology and packet-dispersion technology.

Ipv6 bandwidth measurement generally obtains the available bandwidth of the link through the link busy or idle state. The detection end can actively send a detection message with an IPv6 timestamp extension header to the network, and record the current time of the router hop by hop [40]. After the detection result is obtained, it can be processed by the link-state statistical algorithm to eliminate the deviation caused by processing delay, propagation delay and router clock distortion.

5.1   Single Packet Technology

The single-packet technology is also commonly referred to as a VPS (Variable Packet Size) technology. Its basic idea is to suppose a network path composed of n links (as shown in Fig. 9), send measurement packets of different sizes from the source node to the destination node, and obtain the difference between each node in this path and the source node Time delay [41]. This delay can be OWD (One Way Delay) or RTT (Round Trip Time). After analyzing the delay data, the capacity of each hop link can be obtained.

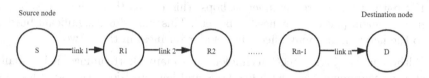

Fig. 9. Single group technology measurement model [39]

The implementation of a single packet is easy, and the measurement work can be completed without the cooperation of source and destination nodes. But the disadvantage is also pronounced, first of all, it must meet the following basic assumptions:

- The transmission time of a data packet is directly proportional to the packet size and inversely proportional to the link capacity.
- Routers are all store and forward.
- All links are single-channel.
- No other network traffic on the transmission path causes measurement packets to be queued.

However, the first and the fourth assumption are very difficult to achieve in real life. In fact, the delay obtained in the measurement process should consist of four parts: Processing Delay, Queuing Delay, Transmission Delay and Propagation Delay. Processing delay refers to the time required for the router to inspect the packet header and decide where to direct the packet. Queuing delay refers to the waiting time experienced by packets after entering the transmission queue. Transmission delay refers to the time required for the router to push all bits of the packet to the link. Propagation delay refers to the time required for a bit to propagate from one router to another. Only the transmission delay is related to the link capacity, and the existence of the other three delays will have an impact on the available packet technology [7]. In addition, in the actual network environment, background traffic always exists, and network managers or operators cannot interrupt various services for measuring bandwidth.

In addition to the above reasons, single packet technology still has the following problems:

- If the measurement adopts OWD, then both the transmitter and receiver need to synchronize the clocks; otherwise, the accurate time delay will not be obtained.
- If RTT is used for the measurement, during the measurement process, the intermediate router may temporarily change the route of the measurement packet due to some special reasons, such as congestion, load balancing, etc., which may cause the forward link and the reverse link asymmetry, which increases the possibility of queuing measurement packets, and ultimately leads to increased measurement errors.
- To find the narrowest link, it is necessary to measure each link one by one. For paths with a large number of hops, this means that a large number of measurement packets may need to be sent. This will bring a significant burden to the network and may affect the regular business in the network.

Although the single packet technology has many shortcomings, it is usually used in measurement tasks with few hops and low precision because of its simplicity.

5.2   Packet-Dispersion Technology

Packet dispersion technology estimates the network path bandwidth by measuring the time interval when the packet arrives at the receiver. According to the different measurement indicators, it can be divided into packet-pair technology and packet-chain technology [42]. The former is mainly used to estimate the path capacity, and the latter is often used to estimate the available bandwidth, that is, the maximum transmission rate that the network path can provide. This section mainly introduces packet-pair.

Compared with the single packet technology, the most significant advantage of the packet pair technology is that it does not need to consider the propagation delay of the link, because, in the absence of background traffic interference, the propagation delay is the same for all measurement packets [43]. Also, the packet pair technology can measure the path capacity value of the entire path every time two packets are successfully sent, while the single packet technology can only measure the bandwidth of one link every time a packet is successfully sent. So, when there are a large number of links on the path to be tested, the efficiency of the packet pair technology is higher than that of the single packet technology.

However, the biggest weakness of grouping technology is that the background traffic existing in the virtual network will have a massive impact on the measurement, especially when the link utilization is high, the error will be unusually large. If the two test packets originally sent back-to-back are inserted in the middle of the transmission process, the time interval between the two packets arriving at the destination will be longer than usual, resulting in the underestimation of the path capacity. This is time extension problem. If the previous packet in the packet pair is blocked by the background traffic packet in front of it, the time interval between the two packets to reach the target end will be shorter than usual, resulting in an overestimation of the bottleneck bandwidth, which is the problem of time compression.

Table 2. Table of main bandwidth measurement tools

Name	Measurement parameters	Technology	Send
Pathchar	Link bandwidth	Single packet	UDP;ICMP
Clink	Link bandwidth	Single packet	UDP;ICMP
Bprobe	Bottleneck bandwidth	Packet-dispersion	ICMP
Nettimer	Bottleneck bandwidth	Packet-dispersion	UDP

6   Open Issues and Future Directions

In this section, we discuss some open issues and research challenges that need to be studied and addressed.

6.1   Open Issues

Limitations of PMTU Single Point Detection Method: If there are multiple tunnels with the same MTU on a path or a tunnel with a larger MTU value behind one tunnel, this method can only find the tunnel closest to the survey point [44].

Alias Address Problem: On the one hand, in terms of detecting the number of IPv6 addresses, the alias problem is not big. Addresses that can be scanned by ICMPv6 or even TCP80 can be regarded as surviving addresses [14]. However, on the other hand, the alias prefix will affect the universality of the detection algorithm based on the seed address, and the surviving addresses detected by the algorithm are likely to be the addresses of the same interface on the network or the same intermediate device response address. The detection algorithm based on the seed address is highly sensitive to the seed address [45]. If the alias problem cannot be adequately handled, the number of real IPv6 devices discovered will be greatly reduced.

Ability to Deal with Randomization: In addition to IPv6's last 64-bit allocation strategy, there are many random situations. With the continuous increase of the IPv6 seed address set, its overall performance characteristics will inevitably show randomness (assuming that all address spaces are used as seeds address set, each bit of IPv6 has an equal probability of 0 and 1, and Shannon's entropy is all 1) [46]. This will make these methods of mining seed address features invalid regularly, even ultimately.

Detection Methods based on Standard Protocols are Restricted: Under IPv4, the activity of the devices in the subnet is mainly detected by exhaustive Ping method. A class C subnet has 256 addresses to be detected. If the asynchronous method is used, it only takes a few seconds to complete the scan, which will not bring the network too much burdensome. The number of nodes in the IPv6 subnet is quite large, and exhaustive detection hosts need to send about 200 billion ICMP response requests, which is unacceptable for any network.

6.2   Potential Areas for Future Research

Heuristic with Feedback Methods: Most working algorithms are based on seed address modelling and use the model to generate addresses, and then use the generated addresses as the input of the scanner. These three parts are often independent, but 6Gen gives people a perfect inspiration that scanning and address generation can feedback to each other [47]. As a result, we can further think about whether feedback can be applied to the modelling process, and whether more optimized heuristic methods with feedback can be proposed.

Improve the Topology System: The vast IP address space will make the algorithm face enormous challenges. The currently implemented IPv6 topology discovery system still has many shortcomings. The design is relatively rough and needs to be further improved, such as anonymous routers and redundant traffic.

Application of Machine Learning Methods [48]: Bayesian models and clustering algorithms are all successfully applied in IPv6 active measurement. In addition, whether some mainstream machine learning models in recent years can be further used in the field of IPv6 measurement is still worthwhile to explore the problem.

7   Conclusion

This paper analyzes the basic principles of IPv6 network measurement, compares classic measurement models and algorithms, and systematically analyzes the new features of the IPv6 protocol. A comprehensive overview of the active measurement methods of IPv6 network in terms of tunnel discovery, address detection, topology discovery, and bandwidth measurement is presented. By analyzing the advantages and disadvantages of classic active measurement technology, we can lay a good foundation for subsequent research on network active measurement.

Acknowledgement. This work is supported by The National Key Research and Development Program of China (2018YFB1800202, 2018YFB0204301) and The National Natural Science Foundation of China (62072465).

References

1. Ao, N., Chen, C.: Measurement study of ipv6 users on private bt. In: Proceedings of the International Conference on Materials Engineering for Advanced Technologies. Trans Tech Publications, pp. 726–731 (2011)
2. Bajpai, V., Schonwalder, J.: Measuring the effects of happy eyeballs. In: Proceedings of the ACM, IRTF and ISOC Applied Networking Research Workshop. Association for Computing Machinery, Inc. (2016)
3. Berger, A., Weaver, N., Beverly, R., Campbell, L.: Internet nameserver IPv4 and IPv6 address relationships. In: Proceedings of.the 13th ACM Internet Measurement Conference (IMC). Association for Computing Machinery (2013)

4. Beverly, R., Berger, A.: Server siblings: identifying shared IPv4/IPv6 infrastructure via active fingerprinting. In: Mirkovic, J., Liu, Y. (eds.) PAM 2015. LNCS, vol. 8995, pp. 149–161. Springer, Cham (2015). https://doi.org/10.1007/978-3-319-15509-8_12

5. Beverly, R., Brinkmeyer, W., Luckie, M., Rohrer, J.P.: IPv6 alias resolution via induced fragmentation. In: Roughan, M., Chang, R. (eds.) PAM 2013. LNCS, vol. 7799, pp. 155–165. Springer, Heidelberg (2013). https://doi.org/10.1007/978-3-642-36516-4_16

6. Claffy, K.C.: Tracking IPv6 evolution: data we have and data we need. ACM SIGCOMM Comput. Commun. Rev. **41**(3), 43–48 (2011)

7. Dainotti, A., et al.: Estimating internet address space usage through passive measurements. ACM SIGCOMM Comput. Commun. Rev. **44**(1), 43–49 (2014)

8. Dhamdhere, A., Luckie, M., Huffaker, B., Claffy, K., Elmokashfi, A., Aben, E.: Measuring the deployment of ipv6: topology, routing and performance. In: Proceedings of the ACM Internet Measurement Conference (IMC). Association for Computing Machinery (2012)

9. Feng, X., Jun, L., Jia, S.: Security analysis for ipv6 neighbor discovery protocol. In: Proceedings of the 2nd International Symposium on Instrumentation and Measurement, Sensor Network and Automation. IEEE Computer Society (2013)

10. Fiebig, T., Borgolte, K., Hao, S., Kruegel, C., Vigna, G.: Something from nothing (there): collecting global IPv6 datasets from DNS. In: Kaafar, M.A., Uhlig, S., Amann, J. (eds.) PAM 2017. LNCS, vol. 10176, pp. 30–43. Springer, Cham (2017). https://doi.org/10.1007/978-3-319-54328-4_3

11. Fiebig, T., Borgolte, K., Hao, S., Kruegel, C., Vigna, G., Feldmann, Anja: In rDNS we trust: revisiting a common data-source's reliability. In: Beverly, R., Smaragdakis, G., Feldmann, A. (eds.) PAM 2018. LNCS, vol. 10771, pp. 131–145. Springer, Cham (2018). https://doi.org/10.1007/978-3-319-76481-8_10

12. Foremski, P., Plonka, D., Berger, A.: Entropy/IP: uncovering structure in ipv6 addresses. In: Proceedings of the ACM Internet Measurement Conference (IMC). Association for Computing Machinery (2016)

13. Giotsas, V., Luckie, M., Huffaker, B., Claffy, K.: IPv6 AS relationships, cliques, and congruence. In: Mirkovic, J., Liu, Y. (eds.) PAM 2015. LNCS, vol. 8995, pp. 111–122. Springer, Cham (2015). https://doi.org/10.1007/978-3-319-15509-8_9

14. Hendriks, L., De Boer, P.T., Pras, A.: IPv6-specific misconfigurations in the DNS. In: Proceedings of the 13th International Conference on Network and Service Management (CNSM). Institute of Electrical and Electronics Engineers Inc. (2017)

15. Jia, L.: Research and implementation of IPv6 network topology discovery (2012)

16. Jia, S., et al.: Tracking the deployment of IPv6: topology, routing and performance. Comput. Netw. **165**, 106947 (2019)

17. Karir, M., Huston, G., Michaelson, G., Bailey, M.: Understanding IPv6 populations in the wild. In: Roughan, M., Chang, R. (eds.) PAM 2013. LNCS, vol. 7799, pp. 256–259. Springer, Heidelberg (2013). https://doi.org/10.1007/978-3-642-36516-4_27

18. Karpilovsky, E., Gerber, A., Pei, D., Rexford, J., Shaikh, A.: Quantifying the extent of IPv6 deployment. In: Moon, S.B., Teixeira, R., Uhlig, S. (eds.) PAM 2009. LNCS, vol. 5448, pp. 13–22. Springer, Heidelberg (2009). https://doi.org/10.1007/978-3-642-00975-4_2

19. Koo, J., Ahn, S.: AMT6: end-to-end active measurement tool for IPv6 network. In: Yang, L.T., Zhou, X., Zhao, W., Wu, Z., Zhu, Y., Lin, M. (eds.) ICESS 2005. LNCS, vol. 3820, pp. 732–740. Springer, Heidelberg (2005). https://doi.org/10.1007/11599555_70

20. Lee, J., et al.: Automatic configuration of IPv6 tunneling in a dual stack host. In: Gavrilova, M.L., et al. (eds.) ICCSA 2006. LNCS, vol. 3981, pp. 487–494. Springer, Heidelberg (2006). https://doi.org/10.1007/11751588_51

21. Li, F., Pan, T., Yang, J., An, C., Wang, X., Wu, J.: Investigating the prefix-level characteristics: a case study in an ipv6 network. In: Proceedings of the International Conference on Computing, Institute of Electrical and Electronics Engineers Inc, Networking and Communications, pp. 824–829 (2015)

22. Li, F., Yang, J., Wang, X., Pan, T., An, C., Wu, J.: Characteristics analysis at prefix granularity: a case study in an IPv6 network. J. Netw. Comput. Appl. **70**, 156–170 (2016)

23. Li, G.E.A.: Overview of ipv6 address detection technology based on seed address (2019)

24. Li, Q., Qin, T., Guan, X., Zheng, Q.: Exploring flow characteristics in IPv6: a comparative measurement study with IPv4 for traffic monitoring. KSII Trans. Internet Inf. Syst. **8**(4), 1307–1323 (2014)

25. Li, X.F., Wang, Y.Z., Luo, W.M.: A novel IPv6 networks measuring algorithm for locating available bandwidth bottleneck. In: Proceedings of the 9th International Conference on Advanced Communication Technology (ICACT). Institute of Electrical and Electronics Engineers Inc. (2014)

26. Liang, J., Jiang, J., Duan, H., Li, K., Wu, J.: Measuring query latency of top level DNS servers. In: Roughan, M., Chang, R. (eds.) PAM 2013. LNCS, vol. 7799, pp. 145–154. Springer, Heidelberg (2013). https://doi.org/10.1007/978-3-642-36516-4_15

27. Liu, Z.: 6Tree: efficient dynamic discovery of active addresses in the IPv6 address space. Comput. Netw. **155**, 31–46 (2019)

28. Loreti, P., Mayer, A., Lungaroni, P., Salsano, S., Gandhi, R., Filsfils, C.: Implementation of accurate per-flow packet loss monitoring in segment routing over IPv6 networks. In: Proceedings of the 21st IEEE International Conference on High Performance Switching and Routing (HPSR). IEEE Computer Society (2020)

29. Lutu, A., Bagnulo, M., Pelsser, C., Maennel, O.: Understanding the reachability of IPv6 limited visibility prefixes. In: Faloutsos, M., Kuzmanovic, A. (eds.) PAM 2014. LNCS, vol. 8362, pp. 163–172. Springer, Cham (2014). https://doi.org/10.1007/978-3-319-04918-2_16

30. Malone, D.: Observations of IPv6 addresses. In: Claypool, M., Uhlig, S. (eds.) PAM 2008. LNCS, vol. 4979, pp. 21–30. Springer, Heidelberg (2008). https://doi.org/10.1007/978-3-540-79232-1_3

31. Marx, M., Schwarz, M., Blochberger, M., Wille, F., Federrath, H.: Context-aware IPv6 address hopping. In: Zhou, J., Luo, X., Shen, Q., Xu, Z. (eds.) ICICS 2019. LNCS, vol. 11999, pp. 539–554. Springer, Cham (2020). https://doi.org/10.1007/978-3-030-41579-2_31

32. Michaelson, G., Huston, G.: Experience with large-scale end user measurement techniques. In 2014 Australasian Telecommunication Networks and Applications Conference (ATNAC), pp. 1–5. IEEE (2014)

33. Padmanabhan, R., Li, Z., Levin, D., Spring, N.: UAv6: alias resolution in IPv6 using unused addresses. In: Mirkovic, J., Liu, Y. (eds.) PAM 2015. LNCS, vol. 8995, pp. 136–148. Springer, Cham (2015). https://doi.org/10.1007/978-3-319-15509-8_11

34. Pezaros, D.P., Hutchison, D., Gardner, R.D., Garcia, F.J., Sventek, J.S.: Inline measurements: a native measurement technique for IPv6 networks. In: Proceedings of the 2004 International Networking and Communication Conference (INCC). Institute of Electrical and Electronics Engineers Inc. (2004)

35. Pezaros, D.P., Hoerdt, M., Hutchison, D.: Low-overhead end-to-end performance measurement for next generation networks. IEEE Trans. Netw. Serv. Manage. **8**(1), 1–14 (2011)
36. Plonka, D., Barford, P.: Assessing performance of internet services on IPv6. In: Proceedings of the 18th IEEE Symposium on Computers and Communications (ISCC). Institute of Electrical and Electronics Engineers Inc. (2013)
37. Plonka, D., Berger, A.: Temporal and spatial classification of active IPv6 addresses. In: Proceedings of the ACM Internet Measurement Conference (IMC). Association for Computing Machinery (2015)
38. Pujol, E., Richter, P., Feldmann, A.: Understanding the share of IPv6 traffic in a dual-stack ISP. In: Kaafar, M.A., Uhlig, S., Amann, J. (eds.) PAM 2017. LNCS, vol. 10176, pp. 3–16. Springer, Cham (2017). https://doi.org/10.1007/978-3-319-54328-4_1
39. Quanjie, Q.: Research on technology upgrade and performance measurement of campus network based on IPv6 (2014)
40. Rohrer, J., LaFever, B., Beverly, R.: Empirical study of router IPv6 interface address distributions. IEEE Internet Comput. **20**(4), 36–45 (2016)
41. Rye, E.C., Beverly, R.: Discovering the IPv6 network periphery. In: Sperotto, A., Dainotti, A., Stiller, B. (eds.) PAM 2020. LNCS, vol. 12048, pp. 3–18. Springer, Cham (2020). https://doi.org/10.1007/978-3-030-44081-7_1
42. Sarrar, N., Maier, G., Ager, B., Sommer, R., Uhlig, S.: Investigating IPv6 Traffic. In: Taft, N., Ricciato, F. (eds.) PAM 2012. LNCS, vol. 7192, pp. 11–20. Springer, Heidelberg (2012). https://doi.org/10.1007/978-3-642-28537-0_2
43. Scheitle, Q., Gasser, O., Rouhi, M., Carle, G.: Large-scale classification of IPv6-IPv4 siblings with variable clock skew. In: Proceedings of the 1st Network Traffic Measurement and Analysis Conference. Institute of Electrical and Electronics Engineers Inc. (2016)
44. Shah, J.L., Parvez, J.: Impact of IPSEC on real time applications in IPv6 and 6to4 tunneled migration network. In: Proceedings of the 2nd IEEE International Conference on Innovations in Information, Embedded and Communication Systems. Institute of Electrical and Electronics Engineers Inc. (2015)
45. Soumyalatha, N., Ambhati, R.K., Kounte, M.R.: IPv6-based network performance metrics using active measurements. In: Chakravarthi, V., Shirur, Y., Prasad, R. (eds.) Proceedings of International Conference on VLSI, Communication, Advanced Devices, Signals & Systems and Networking (VCASAN-2013). Lecture Notes in Electrical Engineering, vol 258. Springer, India (2013). https://doi.org/10.1007/978-81-322-1524-0_53
46. Wullink, M., Moura, G.C.M., Hesselman, C.: Dmap: automating domain name ecosystem measurements and applications. In: Proceedings of the 2nd Network Traffic Measurement and Analysis Conference (TMA). Institute of Electrical and Electronics Engineers Inc. (2018)
47. Xu, T., Zhang, D., Bi, J., Wu, G.: DNS measurement based on IPv6. Jisuanji Gongcheng/Comput. Eng. **32**(8), 144–146 (2006)
48. Yang, L., Li, Z., Zhang, D., Xie, G.: Topology discovery with smallest-redundancy in ipv6. Jisuanji Yanjiu yu Fazhan/Comput. Res. Dev. **44**(6), 939–946 (2007)
49. Xiong, Y., Zhu, P., Liu, Z.: EPLC: an efficient privacy-preserving line-loss calculation scheme for residential areas of smart grid. Secur. Commun. Netw. (2019)
50. Yong, L.: Application of IPv6 tunnel technology in campus network (2013)

Deriving Security Protocols Based on Protocol Derivation System

Ke Yang⑩, Meihua Xiao(✉)⑩, Zifan Song⑩, and Ri Ouyang⑩

School of Software, East China Jiaotong University,
Nanchang 330013, People's Republic of China
xiaomh@ecjtu.edu.cn

Abstract. Protocol Derivation System (PDS) supports syntactic derivations of complex protocols that use cryptographic primitives. However, the PDS is only applicable for two-party interaction protocols, which does not involve in the presence of a Trusted Third Party. In this paper, we proposed an extended PDS that can support key agreement protocols using a Trusted Thirty Party by adding some components, refinements, transformations and removing redundancy rules. A flow chart of deriving security protocols based on the extended PDS is given. The flow chart consists of two layers, the first layer is to get a raw protocol using components, refinements, transformations, the second layer is to remove superfluous protocol steps and improve protocol efficiency by removing redundancy rules. Finally, we get the AOR protocol as an example to illustrate how to derive a security protocol based on the extended PDS.

Keywords: Protocol Derivation System · Security protocols · Trusted third party · Key agreement protocols · Components

1 Introduction

Security protocols are notoriously known to be difficult to get right. Designing error-free security protocols that are impervious to attack techniques, such as freshness and interleaving sessions (i.e. impersonation attacks, man-in-the-middle attacks, oracle attacks, multiplicity attacks and other types of parallel session attacks) is an extremely challenging task [1, 2]. Many protocols proposed in the literature and many protocols exploited in practice turned out to be flawed, or their well-functioning was found to be based on implicit assumptions. Such as the Needham-Schroeder authentication protocol, initiated a large body of work on the design and analysis of cryptographic protocols. In 1995, Gavin Lowe published an attack on the protocol that had apparently been undiscovered for the previous 17 years [3].

However, for security protocols, most of them are designed to meet specific application environment according to actual needs based on the intuition and experience of researchers. It turns out that designing security protocols based on experience and other informal methods has security vulnerabilities.

© Springer Nature Singapore Pte Ltd. 2021
K. He et al. (Eds.): NCTCS 2020, CCIS 1352, pp. 172–184, 2021.
https://doi.org/10.1007/978-981-16-1877-2_12

Formal methods have the characteristics of concise analysis process, and is recognized as an effective way to analyze the security of network protocols [4, 5]. For this reason, researchers have been working on the use of formal verification techniques to analyze the vulnerability of security protocols. A large number of formal methods have been proposed, which including modal logic, theorem proving and model checking [6–8].

In 2005, a framework (denoted DDMP framework) consisting of Protocol Derivation System (PDS) and Protocol Composition Logic (PCL) [9, 10] has been proposed by Datta et al. PDS supports syntactic derivations of complex protocols, starting from basic components, and combining or extending them using a sequence of composition, refinement, and transformation operations [11]. PCL is a Floyd-Hoare style logic that supports axiomatic proofs of protocol properties. The eventual goal is to develop proof methods for PCL for every derivation operation in PDS, thereby enabling the parallel development of protocols and their security proofs. Since the DDMP framework was proposed, this method has been successfully applied to the formal analysis and security proof of many protocols. Datta et al. described the deduction process of various protocols in STS family, and successfully analyzed ISO-9798-3. Roy et al. [12] extended the original PCL for the confidentiality of security protocol. Mitchell et al. [13] have successfully analyzed industrial protocols such as SSL/TLS and Kerberos V5 using PCL. Li et al. [14] proposed a PCL secure user authentication protocol named UCAP for cloud computing. Zhang Junwei et al. [15] proposed a PCL secure and efficient group authentication protocol named TSNP. PCL is an effective way to provide formal proofs to the correctness of security protocols.

There is no doubt that DDMP framework has been successfully applied to a number of industrial network security protocols. However, there are only a little improvement on the PCL rather than PDS after the DDMP framework.

PDS for deriving security protocols consists of a set of basic building blocks named components and a set of operations for constructing new protocols from old ones. These operations can be divided into three different types: composition, refinement and transformation. A component is used as a building block for larger protocols. For example, the Diffie-Hellman key exchange and challenge-response component are basic components providing the desired security properties. The composition operation combines two sub-protocols. Parallel composition and sequential composition are two composition operations. The refinement operation acts on message components of a single protocol. The original PDS mainly aims at the design of two-party interaction protocols including initiator and responder, without the existence of a Trusted Third Party. In addition, the main set of operations for PDS is public key encryption [16]. Although Zhang Junwei et al. [17–19] added some new components, refinements and transformation to the PDS, their work mainly aims to derive the key exchange protocols in the Needham-Schroeder family. Therefore, in order to derive key agreement protocols using a Trusted Third Party in symmetric encryption environment, we intend to extend the PDS.

Our Contributions. In this paper, we study the extension of PDS to support the derivation of key agreement protocols using a Trusted Third Party in symmetric encryption environment.

1. Some components, refinements, transformations and removing redundancy rules are added to the PDS. The extended PDS can carry out protocol design using combination method based on symmetric encryption environment.
2. A flow chart of deriving security protocols based on the extended PDS is given. The flow chart consists of two layers, the first layer is to get a raw protocol using components, refinements, transformations, the second layer is to remove superfluous protocol steps and improve protocol efficiency.
3. Deriving a key agreement protocol in symmetric environment is given as an example to illustrate how the extended PDS supports the derivation of security protocols.

The rest of this paper is organized as follows. Section 2 introduces some preliminaries about PDS and the defects of PDS. Section 3 gives the extension of PDS. In Sect. 4, we give the detailed derivation of security protocols to illustrate how to derive protocols based on the extended PDS. Section 5 concludes.

2 Protocol Derivation System and Its Defects

PDS is a useful method to design and generate security protocols syntactically, which consists of a set of basic building blocks called components and a set of operations for constructing new protocols from old ones. First, we give brief introduction about components, operations and transformations, then the defects in PDS are presented.

2.1 The Explanation of Symbols

A, B, S: Initiator, Responder and Server (Trusted Thirty Party).
K: shared key. In symmetric encryption environment, the encryption key is the same as the decryption key, which can be written as $K = K^{-1}$.
N_a, N_b: random numbers generated by A and B.
m: A message.
K_{xy}:   the shared key K_{xy} between X and Y.
$SIG_X(m)$:   A signature of m created by X.
$X \to Y$: X sends a message m to Y.
$p \Rightarrow q$:   block p is replaced by block q.

2.2 Components

A protocol component consists of a set of roles (e.g., initiator, responder, server), where each role has a sequence of inputs, outputs and protocol actions. Diffie-Hellman key exchange and signature-based authenticator are basic components in PDS.

Diffie-Hellman Component, C_1

The Diffie-Hellman protocol [20] provides a way for two parties to set up a shared key (g^{ir}) which a passive attacker cannot recover. There is no authentication guarantee: the secret is shared between two parties, but neither can be sure of the identity of the other. The initiator and responder role actions are given below.

I: generates random value i and computes g^i (for previously agreed base b)
R: generates random value r and computes g^r (for previously agreed base b)

In this component no messages are sent.

Signature-Based Authenticator, C_2

The signature-based challenge-response protocol shown below is a standard mechanism for one-way authentication.

$I \rightarrow R:$ m
$R \rightarrow I:$ $SIG_r(m)$

It is assumed that m is a fresh value or nonce and that the initiator, I, possesses the public key certificate of responder, R, and can therefore verify the signature.

2.3   Composition

The composition operation used is sequential composition of two protocol components with term substitution. The roles of the composed protocol have the following structure: the input sequence is the same as that of the first component and the output is the same as that of the second component; the actions are obtained by concatenating the actions of the first component with those of the second (sequential composition) with an appropriate term substitution-the outputs of the first component are substituted for the inputs of the second.

2.4   Refinements

Refinement R_1 $SIG_X(m) \Rightarrow EK(SIG_X(m))$, where K is a key shared with the peer. The purpose of this refinement is to provide identity protection against passive attackers

Refinement R_2 $SIG_X(m) \Rightarrow SIG_X(HMAC_K(mID_X))$, where K is a key shared with the peer. While the signature by itself proves that this term was generated by X, the keyed hash in addition proves that X possesses the key K.

Refinement R_3 $SIG_X(m) \Rightarrow SIG_X(m), HMACK(m, ID_X)$, where K is a key shared with the peer. This refinement serves the same purpose as R_2.

Refinement $R4$ $SIG_X(m) \Rightarrow SIG_X(mID_Y)$, where Y is the peer. It is assumed that X possesses the requisite identifying information for Y, e.g., Y's public key certificate, before the protocol is executed. This assumption can be discharged if X receives Y's identity in an earlier message of the protocol.

Refinement $R5$ $g^x \Rightarrow g^x, n^x$, where n^x is a fresh value. The use of nonce enables the reuse of exponentials across multiple sessions resulting in a more efficient protocol.

Refinement R_6 $SIG_X(m) \Rightarrow SIG_X(m), ID_X$, where ID_X denotes the public key certificate of X. This refinement enables that the principals possess each other's public key certificates before the session.

Refinement R7 $E_K(m) \Rightarrow E_K(m), HMAC_{K'}(role, E_K(m))$, where K and K' are keys shared with the peer and role identifies the protocol role in which this term was produced (initiator or responder).

2.5 Transformations

Message Component Move, T_1

This transformation moves a top-level field t of a message m to an earlier message m', where m and m' have the same sender and receiver, and if t does not contain any data freshly generated or received between the two messages. One reason for using this transformation is to reduce the total number of messages in the protocol.

Binding, T_2

Binding transformations generally add information from one part of a protocol to another in order to "bind" the two parts in some meaningful way. The specific instance of this general concept that we use here adds a nonce from an earlier message into the signed portion of a later message, this operation is illustrated as follows.

$$
\begin{array}{lll}
I \rightarrow R: m & & I \rightarrow R: m \\
R \rightarrow I: n, SIG_R\,(m) & \Rightarrow & R \rightarrow I: n, SIG_R\,(m, n) \\
I \rightarrow R: SIG_I\,(n) & & I \rightarrow R: SIG_I\,(m, n)
\end{array}
$$

Cookie, T_3

The purpose of the cookie transformation is to make a protocol resistant to blind Denial-of-Service (DoS) attacks [21]. Under certain assumptions, it guarantees that the responder does not have to create state or perform expensive computation before a round-trip communication is established with the initiator.

$$
\begin{array}{lll}
I \rightarrow R: m_1 & & I \rightarrow R: m_1 \\
R \rightarrow I: m_2 & \Rightarrow & R \rightarrow I: m_2^c, HMAC_{HKR}\,(m_1, m_2^c) \\
I \rightarrow R: m_3 & & I \rightarrow R: m_3, m_1, m_2^c, HMC_{HKR}\,(m_1, m_2^c) \\
& & R \rightarrow I: m_2^e
\end{array}
$$

To sum up, key agreement protocols using a Trusted Thirty Party differ from the STS family and Needham-Schroeder family in the following characteristics.

1. Involvement of a Trusted Third Party. There are only the Diffie-Hellman component and signature-based authenticator component in PDS, which can't derive key agreement protocols using a Trusted Thirty Party
2. Symmetric encryption. Due to the lack of components and operations to describe this kind of encryption type in PDS, symmetric key exchange protocol can't be derived by PDS.
3. More interactive steps. Because of the involvement of Trusted Third Party and symmetric encryption-based authentication, it's necessary to remove superfluous protocol steps and improve protocol efficiency by removing redundancy rules.

3 The Extension for PDS

For the original PDS exists some defects, an extended PDS is proposed. Not only some components and transformations are added in the extended PDS, but also some removing redundancy rules which don't exist in original PDS are proposed, the purpose of these rules is to reduce superfluous protocol steps and improve efficiency.

3.1 Extended Components

Timestamp Component, C_3

$$A \rightarrow S:\ \{A, B\}K_{as}$$
$$S \rightarrow A:\ (A, M, T_S)K_{as}, (B, M, T_S)K_{bs}$$

The client requests from the server. If the request information is encrypted and intercepted by a third party to the request package, the request package can be used for repeated requests. If the server does not prevent replay attack, the server pressure will increase, and the use of timestamp can solve this problem.

Trusted Thirty Party Component, C_4

$$A \rightarrow S:\ A, B$$
$$S \rightarrow A:\ K_{ab}$$
$$A \rightarrow B:\ K_{ab}$$

This component is a basic framework for key agreement protocols with Trusted Thirty Party. In this component, the initiator A sends his identity and B's identity to the server. Then the server responds with shared key K_{ab} between A and B. Finally, the initiator sends K_{ab} to B. This component can provide key agreement but don't ensure authentication and confidentiality.

Challenge-Response Component, C_5

$$C \rightarrow R:\ N_C$$
$$R \rightarrow C:\ E_K \{m, N_C\}$$

In contrast to signature-based authenticator component C_2, this component is encrypted by symmetric key and the request information is a random number rather than message m. Using a series of Challenge-Response steps, which can provide user authentication to some degree.

Three-Party Authentication Component, C_6

$A \rightarrow B$: $A, \{A, N_a\} K_{as}$
$B \rightarrow S$: $\{N_b, B\}K_{bs}, \{A, N_a\} K_{as}$
$S \rightarrow B$: $\{Na, N_b\}K_{bs}$

The three-party authentication component C_6 is a basic framework for key agreement protocols using a Trusted Third Party. Because of the existence of N_a and N_b, This Component can ensure the authentication between A and B through a Trusted Thirty Party.

3.2 Extended Refinements

Refinements are mainly for the cryptographic operation of messages in the protocol. Without changing the number of messages or the structure of the protocol, the necessary security attributes can be added to a message component by refinements.

There are seven refinements R_1–R_7 in the existing PDS system, which are mainly for public key cryptosystem. The cryptographic refinement operations based on symmetric key encryption are as follows.

Refinement R_8: $m \rightarrow E_K (m)$. The message m is encrypted by symmetric key K, no one can decrypt message $E_K (m)$ unless the agent discloses its own private key. This refinement can guarantee the confidentiality of message m to some degree.
Refinement R_9: $E_K(m) \rightarrow E_K(m, X)$. The identity of X is added to $E_K (m)$, therefore this refinement can prevent man-in-middle attack effectively.
Refinement R_{10}: $E_K\{N_a\} \rightarrow E_K(N_a-1)$. Responding the message $E_K(N_a-1)$ rather than $E_K\{N_a\}$, which can refresh the random number N_a and prevent reply attack effectively.
Refinement R_{11}: $E_K(X) \rightarrow E_K(X, N_a)$. By adding the random number N_a, this refinement can prevent reply attack to some degree because of the freshness of N_a

3.3 Extended Transformations

Symmetry, T_8

$A{\rightarrow}B$: $X, A, B, \{N_a, A\}K_{as}$ $\Rightarrow$ $A{\rightarrow}B$: $X, A, B, \{N_a, A, B\}K_{as}$
$B{\rightarrow}S$: $X, A, B, \{N_a, A\}K_{as}, \{N_b, B\}K_{bs}$ $B{\rightarrow}S$: $X, A, B, \{Na, A\}Kas, \{N_b, B\}K_{bs}$

Changing Order, T_9

$A{\to}S: A, B$ $A{\to}B: A$

$S{\to}A:\{K_{ab}, B\}K_{as}, \{K_{ab}, A\}K_{bs}$ $\Rightarrow$ $B{\to}A: A, B$

$A{\to}B:\{K_{ab}, A\}K_b$ $S{\to}B: \{K_{ab}, A\}K_{bs}, \{K_{ab}, B\}K_{as}$

$B{\to}A: \{K_{ab}, B\}K_{as}$

Index, T_{10}

To denote a run of protocol, we add a unique index to every step of transformed protocol.

$A{\to}B: A, B$ $A{\to}B: I, A, B$

$B{\to}S: A, B$ $\Rightarrow$ $B{\to}S: I, A, B$

$S{\to}B: \{K_{ab}\}K_{bs}, \{K_{ab}\}K_{as}$ $S{\to}B: I, \{K_{ab}\}K_{bs}, \{K_{ab}\}K_{as}$

$B{\to}A: \{K_{ab}\}K_{as}$ $B{\to}A: I, \{K_{ab}\}K_{as}$

Consistency, T_{11}

Sometimes we need to ensure the consistency of shared key through key agreement protocols, so the transformation T_{11} is proposed.

$A{\to}B: A, B$ $A{\to}B: A, B$

$B{\to}S: A, B$ $\Rightarrow$ $B{\to}S: A, B$

$S{\to}B: \{K_{ab}\}K_{bs}, \{K_{ab}\}K_{as}$ $S{\to}B: \{K_{ab}\}K_{bs}, \{K_{ab}\}K_{as}$

$B{\to}A: \{K_{ab}\}K_{as}$ $B{\to}A: \{K_{ab}\}K_{as}, \{B\}K_{ab}$

3.4 Removing Redundancy Rules

There are some redundant operations and items in redundancy protocol, which leads to redundant authentication and affects efficiency. At the same time, the operation items in the security protocol are too long, which will affect the efficiency, especially when both agents use public key to achieve security properties. Therefore, we introduce the following removing redundancy rules.

Cascade Connection, RR_1

If there are multiple encryption structures in the protocol step i, and these encryption structures use the same cryptographic algorithm and key, thus these can be combined into one structure. The cascade connection is expressed in logical language as follows:

B receive $E_K(m_1)$

$\Rightarrow$ B receive $E_K(m_1, m_2)$

B receive $E_K(m_2)$

Deleting Duplicate Information, RR_2

If the same message appears twice in a protocol step, only one of them is retained. The deleting duplicate information is expressed in logical language as follows:

$$B \ receive \ E_K(m)$$
$$\Rightarrow \quad B \ receive \ E_K(m)$$
$$B \ receive \ E_K(m)$$

Freshness, RR_3

If the data items used to prove freshness in protocol step i (such as random numbers) always appear at the same time as a data item with freshness in the protocol, then the data items used to prove freshness can be deleted. The rule is expressed in logical language as follows:

$$Has(X, MN) \ \wedge \ Fresh(MN)$$
$$\Rightarrow Has(X, MN) \ \wedge \ Fresh(MN)$$
$$Has(X, M) \ \wedge \ Fresh(M)$$

Plaintext, RR_4

If a plaintext m appears in the encryption part of the protocol, the plaintext m is encrypted first and then simplified by cascade connection RR_1.

4 Deriving Security Protocols Based on the Extended PDS

Based on the extended PDS, we give a flow chart to derive security protocols. This flow chart can be divided into two layers. The first layer is to get a raw protocol meeting expected security properties. First, combining the basic cryptographic primitives, we can get some components, which are the basic framework for different kinds of security protocols. According to expected security properties, for example, the mutual authentication or the key agreement between initiator and responder, choosing some correct components and then getting a single step protocol. Next, based on the composition rules, appropriate refinements and transformations, a raw protocol can be concluded. The second layer is to remove redundancy and improve efficiency. Through removing redundancy rules, a target protocol can be derived. The above steps are illustrated in Fig. 1.

In order to better illustrate how to derive a security protocol based on the extended PDS, we will derive a three-party key agreement protocol as a case study.

1. First, our objective protocol is a three-party key agreement protocol, so we choose C_4 as the basic component, which provides key agreement using a Trusted Thirty Party.
2. Protocol P_1 obtained by applying R_8 to C_4.

$$A \rightarrow S: \quad A, B$$

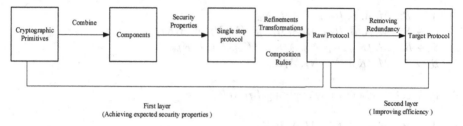

Fig. 1. The flow chart of deriving security protocols based on the extended PDS

$S \rightarrow A$: $\{K_{ab}\}K_{as}$, $\{K_{ab}\}K_{bs}$
$A \rightarrow B$: $\{K_{ab}\}K_{bs}$

3. Protocol P_2 obtained by applying T_9 to P_1.

$A \rightarrow B$: A
$B \rightarrow S$: A, B
$S \rightarrow B$: $\{K_{ab}\}K_{as}$, $\{K_{ab}\}K_{bs}$
$B \rightarrow A$: $\{K_{ab}\}K_{as}$

4. Protocol P_3 obtained by applying T_{10} to P_2.

$A \rightarrow B$: M, A
$B \rightarrow S$: M, A, B
$S \rightarrow B$: $M, \{K_{ab}\}K_{as}$, $\{K_{ab}\}K_{bs}$
$B \rightarrow A$: $M, \{K_{ab}\}K_{as}$

5. The goal of key agreement can be achieved by protocol P_3 basically. Now we need to add component C_6 which provides mutual authentication between initiator and responder. Protocol P_4 obtained by applying C_6 to P_3.

$A \rightarrow B$: $M, A, \{N_a, A\}K_{as}$
$B \rightarrow S$: $M, A, B, \{N_a, A\}K_{as}$, $\{N_b, B\}K_{bs}$
$S \rightarrow B$: $M, \{K_{ab}\}K_{as}, \{N_a, A\}K_{as}$, $\{K_{ab}\}K_{bs}$, $\{N_b, B\}K_{bs}$
$B \rightarrow A$: $M, \{K_{ab}\}K_{as}, \{N_a, A\}K_{as}$

6. We can learn that some messages are encrypted by the shared key, It's necessary to remove superfluous protocol steps and improve protocol efficiency. So protocol P_5 obtained by applying cascade connection RD_1 to P_4.

$A \rightarrow B$: $M, A, \{N_a, A\}K_{as}$
$B \rightarrow S$: $M, A, B, \{N_a, A\}K_{as}$, $\{N_b, B\}K_{bs}$
$S \rightarrow B$: $M, \{K_{ab}, N_a, A\}K_{as}$, $\{K_{ab}, N_b, B\}K_{bs}$
$B \rightarrow A$: $M, \{K_{ab}, N_a, A\}K_{as}$

7. Protocol P_6 obtained by composition P_5 and T_8.

$A \rightarrow B$: $M, A, B, \{N_a, M, A, B\}K_{as}$
$B \rightarrow S$: $M, A, B, \{N_a, M, A, B\}K_{as}, \{M, N_b, A, B\}K_{bs}$
$S \rightarrow B$: $M, \{K_{ab}, N_a, N_b, A\}K_{as}, \{K_{ab}, N_b, B\}K_{bs}$
$B \rightarrow A$: $M, \{K_{ab}, N_a, N_b, A\}K_{as}$

8. Protocol P_7 obtained by applying T_{11} to P_6.

$A \rightarrow B$: $M, A, B, \{N_a, M, A, B\}K_{as}$
$B \rightarrow S$: $M, A, B, \{N_a, M, A, B\}K_{as}, \{M, N_b, A, B\}K_{bs}$
$S \rightarrow B$: $M, \{K_{ab}, N_a, N_b, A\}K_{as}, \{K_{ab}, N_b, B\}K_{bs}$
$B \rightarrow A$: $M, \{K_{ab}, N_a, N_b, A\}K_{as}, \{N_b\}K_{ab}$

These above steps are shown in a flow chart Fig. 2.

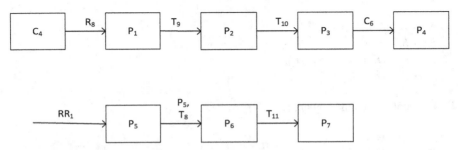

Fig. 2. Deriving three-party key agreement protocol based on the extended PDS

The final protocol P_7 is an AOR protocol, whose security properties have been proved by PCL in reference [22].

Through the combinations and refinements of the basic components, the target security protocol is obtained. Compared with a security protocol based on expert intuition and experience, the target security protocol obtained in this paper is more secure to a certain extent. Of course, to further ensure the security properties the target protocol, we can provide logical proof for the security protocols based on PCL.

In additional, in order to illustrate the advantage of the extended PDS, we give an overall comparison with Datta [9] and Zhang Junwei's methods [17, 18].

	Trusted Third Party	Symmetric Encryption	Removing Redundancy Rules	Timestamp	Three-Party Authentication
Datta	×	×	×	×	×
Zhang Junwei	√	√	×	√	×
This paper	√	√	√	√	√

Compared with Datta and Zhang Junwei's methods, the main advantages of this paper add a Trusted Third Party, removing redundancy rules and three-party authentication component, which support the derivation of public key or symmetric key protocols with trusted third party. In addition, some removing redundancy rules are added in the PDS in order to reduce redundant information and improve efficiency.

5 Conclusions

PDS is a useful method to design and generate security protocols syntactically. However, the PDS is only applicable for two-party interaction protocols, which does not involve in the presence of a Trusted Third Party. In this paper, we proposed an extended PDS that can support key agreement protocols using a Trusted Thirty Party by adding some components, refinements, transformations and removing redundancy rules. A flow chart of deriving security protocols based on the extended PDS is given. The flow chart consists of two layers, the first layer is to get a raw protocol using components, refinements, transformations, the second layer is to remove superfluous protocol steps and improve protocol efficiency by removing redundancy rules. We get the improved version of Otway-Rees protocol as an example to illustrate how to derive a security protocol based on the extended PDS.

In DDMP framework designed by Datta, PCL can provide logical proof for the security protocols generated by PDS. But at present, PCL can't support the logical proof of every step in the PDS system. In additional, deriving different types of security protocols supported by PDS is still our future work.

Acknowledgements. This work is supported by following projects: The National Natural Science Foundation of China under Grant No.61962020 and No.61562026, the Jiangxi Province Key Subject Academic and Technical Leader Funding Project under Grant No. 2017XSDTR0105.

References

1. Sihan, Q.: Twenty years development of security protocols research. J. Software **14**(10), 1740–1752 (2000)
2. Yong, L., Fan, Z., Ming, Z.: Survey on security protocol space information network. Comput. Sci. **44**(04), 202–206 (2017)
3. Lowe, G.: Breaking and fixing the Needham-Schroder public-key protocol using FDR. In: Proceeding of TACAS, LNCS 1055, Spring, pp. 147–166 (1996)
4. Avalle, M., Pironti, A., Sisto, R.: Formal verification of security protocol implementations: a survey. Formal Aspects of Comput. **26**(1), 99–123 (2014)
5. Jian, W., Naijun, Z., Fen, X., Liu, Z.: Overview of formal methods. J. Software, **30**(01), 33–61 (2019)
6. Agray, N., Wiebe, V.D.H., De Vink, E.: On BAN logics for industrial security protocols. In: International Workshop of Central and Eastern Europe on Multi-Agent Systems. Springer, Berlin (2001)
7. Hess, A.V., Sebastian, M.: Formalizing and proving a typing result for security protocols in Isabelle/HOL. In: 2017 IEEE 30th Computer Security Foundations Symposium (CSF). IEEE (2017)
8. Basin, D., Cremers, C., Meadows, C.: Handbook of Model Checking- Model Checking Security Protocols, pp. 727–762. Springer, Cham (2018)

9. Datta, A., Derek, A., Mitchell, J.C., Roy, A.: "Protocol Composition Logic (PCL)", Electronic Notes in Theoretical Computer Science. Gordon D, Plotkin Festschrift (2007)

10. Derek, A.: Formal analysis of security protocols: protocol composition logic, Ph.D. Dissertation, Stanford University (2007)

11. Datta, A., Derek, A., Mitchell, J.C., Pavlovic, D.: A derivation system and compositional logic for security protocols. J. Comput. Security **13**(3), 423–482 (2005)

12. Roy, A., Datta, A., Derek, A., Mitchell, J.C., Seifert, J.-P.: Secrecy analysis in protocol composition logic. In: Formal Logical Methods for System Security and Correctness, IOS Press (2008)

13. Datta, A., Anupam, et al.: Computationally sound compositional logic for key exchange protocols. In: 19th IEEE Computer Security Foundations Workshop (CSFW 2006), IEEE (2006)

14. Li, X., Zhang, J., Ma, J.: UCAP: a PCL secure user authentication protocol in cloud computing. J. Commun. **39**(08), 94–105 (2018)

15. Li, X., Zhang, J., Ma, J., Hai, L.: TSNP: a novel PCL-Secure and efficient group authentication protocol in space information network. J. Comput. Res. Dev. **53**(10), 2376–2392 (2016)

16. Cremers, C.: On the protocol composition logic PCL. In: Proceedings of the 2008 ACM Symposium on Information, Computer and Communications Security (2008)

17. Zhang, J., Yang, C., Ma, J.: Protocol derivation system for the Needham-Schroeder family. In: 2011 6th International ICST Conference on Communications and Networking in China (CHINACOM), pp. 836–840 (2011)

18. Zhang, J., Ma, J., Chao, Y.: Protocol derivation system for the needham-schroeder family. Secur. Commun. Netw. **8**, 2687–2703 (2015)

19. Lu, L.: Study on theory and applications of security protocols formal analysis, Ph.D. Dissertation, Xidian University (2012)

20. Diffie, W., Hellman, M.E.: New directions in cryptography. IEEE Trans. Inf. Theory, IT-**22**(6), 644–654 (1976)

21. Datta, A., Mitchell, J.C., Pavlovic, D.: Derivation of the JFK protocol, Technical Report KES.U.02.03, Kestrel Institute (2002)

22. Lu, L., Duan, X., Ma, J.: Improvement and formal proof on protocol Otway-Rees. J. Commun. **33**(1), 250–254 (2012)

An Offloading Strategy Based on RSU Cooperation for Vehicular Edge Computing System

Qi Fu[✉], Kunhao Shang, Fei Wang, Bing Ba, and Yiping Wen

Hunan University of Science and Technology, Xiangtan 411201, China
fuqi@hnust.edu.cn

Abstract. In the Internet of Vehicles, the delay requirements for real-time applications between vehicles are very high. When the vehicle is driving on the road, it will inevitably exchange tasks by the Road Side Unit (RSU). After the vehicle offloads the calculation task to the RSU, and leave the communication coverage of the RSU, the vehicle will not receive the computing results of the offloading task. The completion time of the task will increase, and this will reduce the safety of the vehicle. In order to reduce the delay of task completion, A communication structure for adjacent RSUs cooperation calculation is proposed, and an algorithm based on the distance between the vehicle and the RSUs is introduced to solve the remaining driving time of the vehicle within the RSU communication coverage. The remaining time is used to screen out messages to control the number of messages forwarded between RSUs. Finally, A test between non-cooperation-RSU and cooperation-RSU scenes is performed and analyzed in OMNet+ +. The results show the strategy proposed in this paper can reduce the task completion time by 25%.

Keywords: RSU collaboration · Message forwarding · Task offloading · Vehicle edge computing · Internet of Vehicles

1 Introduction

Nowadays, the development of smart cars is valued, especially with the development of 5G, car functions become more abundant. It is hoped that cars can realize most of the functions of smart phones, thus bringing convenience to our travel [1]. However, in-vehicle applications and mobile phone applications have different requirements for delay, such as natural language processing, intelligent driving [2] and the implementation of cooperative traffic management systems, some of which require high delays. With the development of cloud computing technology, the development of in-vehicle network has attracted more and more people's attention, and related technologies have slowly begun to mature [3]. Although cloud computing has a lot of computing resources, it will generate high latency. For applications with low latency requirements, cloud computing can be used to provide computing services, but for real-time applications, cloud computing

© Springer Nature Singapore Pte Ltd. 2021
K. He et al. (Eds.): NCTCS 2020, CCIS 1352, pp. 185–200, 2021.
https://doi.org/10.1007/978-981-16-1877-2_13

cannot be used [4]. The vehicle itself has certain computing resources, namely On-board Computers (OBCs), but due to the high hardware cost of on-board computing, increasing on-board computing resources will increase vehicle manufacturing costs [5]. Edge Computing (EC) refers to a technology that allows data processing at the edge of the network and then returns calculation results. For example, smart phones are the edge of wearable networked devices such as watches. The data calculation of these wearable devices can be done by smart phones [6].

For the Internet of Vehicles, edge computing has led to Mobile Edge Computing (MEC) and Vehicle Edge Computing (VEC). Mobile edge computing is proposed to allow applications to run on the edge of the network, and it is characterized by low latency and high bandwidth [7]. For the vehicle network, the edge of the network is the RSU, and the computing resources of the RSU are limited. The use of MEC and cloud computing in VEC can effectively solve the problem of insufficient RSU computing resources. The vehicle uses the online unloading scheduling algorithm to determine whether the computing task needs to be unloaded. If the vehicle decides not to unload the task, the computing task is completed by OBCs. If the vehicle decides to unload the task, it will send the task to the RSU. After receiving it, the RSU decides whether to execute the task locally or send the task to the cloud server for execution [8].

In the VEC architecture, after the vehicle offloads the computing task to the roadside unit R, it leaves the communication coverage area of R before receiving the returned calculation result and reaches the communication coverage area of the next RSU. At this time, because there is no communication between adjacent RSUs, the vehicle cannot receive the R calculation result in time. The vehicle can only wait for the task request to time out before unloading the task to the RSU again. This process will quickly increase the completion time of the task. Vehicles traveling on the road will frequently switch RSUs, and the message delay during the switching process cannot be ignored.

In order to solve this problem, a communication structure is introduced with the cooperation between adjacent RSUs on the basis of the cooperation of the RSU and the cloud server, and the adjacent RSUs are connected together through optical fibers. Since not all calculation results of tasks have to be forwarded to adjacent RSUs, an algorithm is also designed based on the distance between the vehicle and the RSU to solve the remaining travel time of the vehicle in the current RSU communication coverage. The purpose of this algorithm is to screen out messages that need to be forwarded to adjacent RSUs. The algorithm calculates the remaining travel time of the vehicle in the current RSU communication area. Through the remaining travel time of the vehicle, the RSU can determine whether the vehicle can receive the calculation result before leaving its communication area.

The remainder of this paper is organized as follows: In Sect. 2, the related works are introduced. In Sect. 3, the proposed communication structure and algorithm are discussed. In Sect. 4 the simulation is performed and analyzed. In Sect. 5 the paper is concluded.

2   Related Works

Cloud computing technology is relatively mature. In recent years, people have proposed mobile cloud computing in order to extend cloud computing technology to mobile devices. There are also many researches on mobile cloud computing. These studies are dedicated to using different methods to solve the problem of computing offloading. If too many mobile device users offload computing tasks to the cloud through wireless access at the same time, serious interference is likely to occur between them, which will have a great impact on the data transmission rate, thereby increasing the data transmission time. Chen et al. [9] proposedto use game theory methods in mobile cloud computing to perform computing offloading. Users can self-organize into mutually satisfactory computing offloading decisions through games. Zheng et al. [10] showed that mobile computing is a process performed in a dynamic environment and the mobile user offloads the calculation of the wireless channel to randomly change. The article proposes to express the unloading decision process of mobile users in a dynamic environment as a random game.

Although cloud computing has a lot of computing resources, the delay caused by its long-distance transmission is not negligible, and the connection with the cloud becomes unstable due to the mobility of vehicles. In order to solve this problem, edge computing began to get the attention of researchers. Edge computing installs computing resources such as RSU at the edge of the network, thereby reducing the time for large amounts of data transmission, thereby meeting the needs of real-time vehicle applications. The research on edge computing is mainly to reduce the waiting time of computing offloading and maximize the use of edge computing resources. Wang et al. [11] were dedicated to using dynamic voltage regulation technology to jointly optimize communication and computing resources, in order to achieve partial computing offloading to reduce application delay and energy consumption. Tang et al. [12] proposed an optimized framework for offloading from a single mobile device to multiple edge devices. By jointly optimizing the task allocation decision and the frequency of the central processing unit of a single device, the execution delay of the total task and the energy consumption of a single device are minimized.

Although the edge device can quickly process the data generated by the vehicle, it has limited computing resources and is far inferior to cloud computing. When the data to be calculated is relatively large, the computing resources of the edge device will be fully occupied, and other computing needs can only wait. In order to pursue the best interests of the entire system, vehicles cannot arbitrarily apply for the resources they need. Zhang et al. [13] In order to solve the problem of unloading transmission becoming complicated due to the dynamic topology changes caused by the mobility of vehicles, a cloud-based mobile edge computing offloading framework in the Internet of Vehicles was proposed. The framework considered the heterogeneous requirements of computing tasks and the mobility of vehicles. Wan et al. [14] uninstalled the computing applications in the overloaded edge node to other idle edge nodes. The article proposed a multi-objective optimization problem to select a suitable destination edge node. The purpose was to minimize the unloading delay and unloading cost of vehicle applications, and to achieve load balancing of edge nodes. Wang et al. [15] proposed multi-access edge computing. Multi-access edge computing can meet the low latency and high reliability requirements of corresponding programs in the vehicle network. The author added a game method to multi-access edge computing and designed a revenue function to determine the amount of computing resources each vehicle applies for. This article proposes a multi-user non-cooperative computing offloading game. Based on the game model, a distributed optimal corresponding algorithm is constructed to maximize the utility of each vehicle.

Zhao et al. [16] proposed to combine MEC and cloud computing to perform collaborative computing, that is, cloud-assisted mobile edge. Cloud-assisted mobile edge computing is a combination of mobile edge computing and cloud computing. Vehicles can offload their own computing tasks to MEC or cloud servers for computing. By calculating the unloading strategy and resource allocation strategy, the vehicle decides whether to unload the computing task. Wang et al. [8] in order to improve the performance of the vehicle edge system, online offloading scheduling and resource allocation algorithms were proposed. Simulate the process of vehicles requesting computing resources as a game process. In the game, each car will pursue its own best interests. Therefore, each car will calculate the profit when requesting the resource to calculate whether it will benefit from the request for the computing resource, so as to decide whether to request the computing resource. This game process will eventually reach equilibrium, the Nash equilibrium.

3 Proposed Solution

3.1 Cooperation Communication Architecture

This article selects the communication network architecture composed of cloud servers, edge servers and vehicle nodes. In the highway scenario, it is proposed to communicate and cooperate with adjacent edge servers to speed up task completion and reduce delay. Edge server (RSU) uses optical fiber to connect adjacent RSUs to ensure that messages can be forwarded between adjacent RSUs. RSU has certain computing resources and is deployed on the roadside. Vehicles can apply for RSU computing resources to complete their own computing tasks. Each RSU has a fixed communication range r, and the communication coverage ranges of any two RSUs do not overlap. Adjacent RSUs can use the Infrastructure to Infrastructure (I2I) communication protocol to transmit messages. Vehicle nodes will generate computing tasks, which can be completed by OBCs. If the task requires large computing resources, the vehicle can offload the computing tasks to RSU or the cloud to complete. The vehicle and RSU transmit data through the Vehicle to Infrastructure (V2I) communication protocol. The cooperative working process of adjacent RSUs is shown in Fig. 1.

In Fig. 1, when the RSU receives the task of unloading the vehicle, it must determine whether the vehicle can receive the return result before leaving the communication coverage. This can be divided into two cases. The first case is that within the communication range of the roadside unit R, the vehicle offloads the calculation task to R. Before the calculation result is returned, the vehicle leaves the communication coverage area of R and reaches the adjacent RSU, causing the vehicle to fail to receive the returned result. The second case is that the vehicle can receive the return result before it leaves the communication coverage of R. When the first situation occurs, after R calculates the result, it can only discard the result because it cannot find the received vehicle node. However, since the vehicle has not received the result of the task, it can only unload the task again after waiting for a timeout time, which will increase the time it takes to complete the task sharply. This paper uses the method of adjacent RSU cooperation to solve the problem that the vehicle cannot receive the calculation result. The specific method is that R forwards the calculation result to the adjacent RSU, and then the adjacent RSU sends the calculation result to the target vehicle. When the second situation occurs, R sends the calculation result directly to the target vehicle.

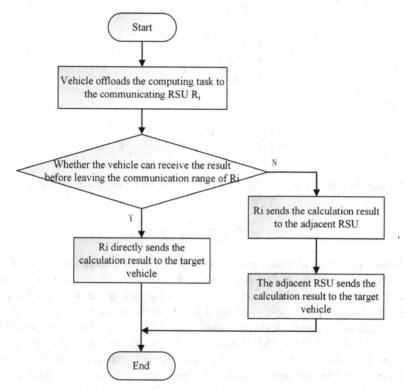

Fig. 1. RSU cooperating communication flow chart.

3.2 Cooperation Calculation Strategy

Resource Queue Model. Each RSU has a limited communication bandwidth. In order to improve its bandwidth resource utilization, we have designed a dynamically calculated message queue for each RSU. In this article, each RSU has designed two dynamic queues, which are used to store messages from vehicles and messages from adjacent RSUs. The message queue mainly depends on the arrival rate of the message task and the computing power of the RSU. In this article, the two dynamic queues in the RSU adopt the M/M/s queuing model, and s is the number of cloud servers or the number of adjacent RSUs. Assuming that the average arrival rate of message m is a Poisson distribution of λ_m. The computing power of RSU is the exponential distribution of μ_i ($i = 1, 2, \ldots, s$). Then the service intensity of the M/M/s system is shown in formula 1.

$$\rho_{mi} = \frac{\lambda_m}{i\mu_i} \tag{1}$$

To keep the system stable, must be less than 1, and the corresponding equilibrium distribution is shown in formula 2.

$$\pi_k = \begin{cases} \dfrac{(i\rho_{mk})^k}{k!}\pi_0, & 0 \leq k \leq i-1 \\ \dfrac{\rho_{mi}i^i}{i!}\pi_0, & k \geq i \end{cases} \tag{2}$$

where

$$\pi_0 = \left[\sum_{k=0}^{i-1} \frac{(k\rho_{mj})^k}{k!} + \frac{i^i \rho_{mi}^i}{i!(1 - \rho_{mi})} \right]^{-1} \tag{3}$$

The waiting time for the message m to be forwarded to the adjacent roadside unit is as in formula 4.

$$t_{im} = \frac{\rho_{mi}}{\lambda_m (1 - \rho_{mi})^2} \pi_i \tag{4}$$

Remaining Time Calculation. In actual situations, not all calculation results of tasks need to be forwarded to adjacent RSUs. Only those calculation results that cannot be received need to be forwarded. In order to filter out the messages that need to be sent to adjacent RSUs, this paper designs an algorithm to calculate the remaining travel time of the vehicle in the current RSU communication coverage area based on the distance between the vehicle and the RSU. According to the calculated remaining travel time, RSU can filter messages that do not have enough time to return the calculation result to the target vehicle, and then forward these messages to the adjacent RSU, and the adjacent RSU will return the calculation results to the target vehicle.

The focus of this algorithm is to obtain the remaining travel distance of the vehicle in the current RSU communication coverage area, and use the ratio of the remaining travel distance to the vehicle speed to calculate the remaining travel time. We define the distance between the vehicle u and the RSU at different times within the communication range of the RSU as $D_{VR}^u = \{D_{VR1}^u, D_{VR2}^u, \cdots, D_{VRn}^u\}$. When the distance between the vehicle and the RSU is shortened, it means that the vehicle is driving in the direction of the communication coverage area of the RSU and is gradually approaching the RSU. When the distance between the vehicle and the RSU increases, it means that the vehicle is driving away from the communication coverage area of the RSU and gradually moves away from the RSU. Set the communication coverage of RSU to a circle with a radius of 500m. Since the road within the coverage of RSU communication is very likely to be curved, it is difficult to directly calculate the actual length of the road. Therefore, when the vehicle is driving away from the RSU communication coverage area, a point on the road can be determined according to the vehicle coordinates, and then the tangent of the road at this point can be calculated. Calculate the remaining travel distance of the vehicle in the RSU communication coverage area according to the tangent, so as to find the remaining travel time. Since the vehicle will frequently send messages to the RSU when the computing task is unloaded, when sending the message, the vehicle will send both the current coordinates and the coordinates of the last message sent to the RSU. The straight line represented by these two points can be approximated as a tangent to the road S

When the vehicle is moving, we use (X^u, Y^u) to represent the coordinates of the vehicle u when sending a message to the RSU this time. Use $(beforeX^u, beforeY^u)$ to record the coordinates of vehicle u when it sent a message to RSU last time. According to (X^u, Y^u) and $(beforeX^u, beforeY^u)$, a linear equation L can be obtained. Since the interval between consecutive messages are very small, L can be regarded as the tangent equation of the current road S at point (X^u, Y^u). Let D_{RS} represent the vertical distance from RSU to this tangent equation. The communication circle of RSU will intercept a part of the tangent line L as the chord, and use L_s to represent half of the chord length. The visual representation of each variable is shown in Fig. 2.

In Fig. 2, the center of the circle represents the coordinates of the roadside unit R, the road S within the coverage of the RSU communication is represented by a curve, and the position of the car represents the coordinates when the vehicle sends a message. The vehicle is driving towards point B along the road. The L_{VRU} represents the distance from the vehicle to the vertical intersection of the tangent line and the RSU, The L_{ru} represents the remaining driving distance of the vehicle u in the RSU communication coverage area. The R_c represents the communication radius of the RSU, the communication radius of the RSU is 500 m and the D_{VR}^u represents the distance between the vehicle and the RSU.

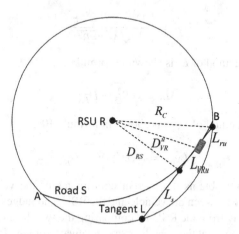

Fig. 2. Visual representation of variables

This algorithm only considers the delay required for the task completion of the vehicle during the RSU switching process. Therefore, this algorithm only needs to calculate the remaining travel time when the vehicle is about to leave the RSU communication coverage area. The calculation method of the remaining travel time of the vehicle in the RSU communication coverage area is shown in formula 5.

$$t_{ru} = \frac{L_{ru}}{V_u} \tag{5}$$

The t_{ru} represents the remaining travel time of vehicle u in the RSU communication coverage area. V_u represents the current speed of the vehicle u.
Where

$$L_{ru} = L_s - L_{VRu} \tag{6}$$

The D_{VRi}^u represents the distance between the current position of the vehicle u and the currently communicating RSU. The D_{VRi-1}^u represents the distance between the previous position of the vehicle u and the currently communicating RSU. Formula 6 indicates that when the distance between the vehicle and the RSU increases, it indicates that the vehicle is driving in a direction away from the communication coverage area of the RSU. At this time, it is necessary to calculate the remaining driving distance of the vehicle within the RSU communication coverage area and calculate the remaining travel time. The calculation method of the distance between the vehicle and the RSU is shown in formula 7, where (X^R, Y^R) represents the coordinates of RSU.

$$D_{VRi}^u = \sqrt{(X^u - X^R)^2 + (Y^u - Y^R)^2} \tag{7}$$

Let the equation of L be $A_x + B_y + C = 0$, Then the calculation method of D_{RS} is shown in formula 8.

$$D_{RS} = \left| \frac{AX^R + BY^R + C}{\sqrt{A^2 + B^2}} \right| \tag{8}$$

The calculation method of L_s is shown in formula 9.

$$L_s = \sqrt{R_C^2 - D_{RS}^2} \tag{9}$$

The calculation method of L_{VRu} is shown in formula 10.

$$L_{VRu} = \sqrt{(D_{VRi}^u)^2 - D_{RS}^2} \tag{10}$$

The algorithm first obtains the current coordinates of the vehicle and the RSU, calculates the distance between the vehicle and the RSU, and judges whether the vehicle is approaching or away from the RSU by comparing the two distances before and after. If it is far away, it means that the vehicle is driving away from the RSU communication coverage area. Then calculate the remaining distance of the vehicle in the RSU, and calculate the remaining travel time by the remaining distance. If it is close, it means that the vehicle is driving toward the center of the RSU communication coverage area, and the vehicle will continue to travel within the RSU communication coverage area for a period of time. At this time, there is no need to calculate the remaining driving distance. The algorithm flow chart is shown as in Fig. 3.

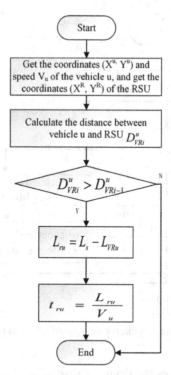

Fig. 3. Algorithm for calculating the remaining travel time of the vehicle

Finally, according to the obtained remaining travel time t_{ru}, it is decided whether the message is sent to the target vehicle or forwarded to the adjacent RSU. We use t^{e2e} to represent the point-to-point delay of the message when the vehicle and the RSU are connected stably. When t_{ru} is greater than t^{e2e}, it means that the target vehicle can receive the returned calculation result before leaving the communication coverage area of this RSU, and the calculation result is directly sent to the target vehicle node.

When t_{ru} is less than t^{e2e}, it means that the vehicle cannot receive the returned calculation result before it leaves the RSU communication coverage area, and then forwards the calculation result to the adjacent RSU, and the adjacent RSU sends the calculation result to the target vehicle. Through the above steps, the messages that need to be forwarded to adjacent RSUs can be screened out, thereby reducing the number of messages forwarded between adjacent RSUs and reducing the bandwidth pressure between adjacent RSUs. The execution process is shown in Fig. 4.

Delay Calculation Method. According to the basic network transmission model, it can be concluded that the communication delay of tasks in the RSU cooperative communication structure includes processing delay, transmission delay, queuing delay and propagation delay. The communication delay calculation method for task m of vehicle u is shown in formula 11.

$$t_{um}^c = t_{proc} + t_{trans} + t_{queue} + t_{prop} + m * t_{prop}' \tag{11}$$

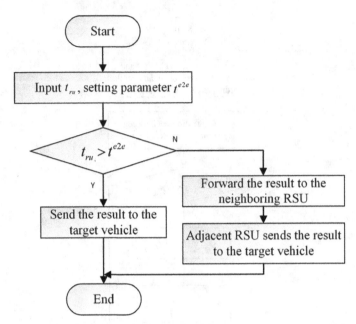

Fig. 4. Algorithm for calculating the remaining travel time of the vehicle

Queuing delay and processing delay are determined by the network environment, and transmission delay is determined by the size and transmission rate of the transmitted packet.The t_{prop} represents the propagation delay from the request sent to the result received within the current RSU communication coverage. t'_{prop} represents the delay caused by the propagation of a message from the current RSU to the adjacent RSU. When the message is transmitted from the current RSU to the adjacent RSU, the value of m is equal to 1, and when the message is only propagated within the current RSU, the value of m is equal to 0.

If the adjacent RSU is not communicating, the vehicle will leave its communication coverage area after sending a task calculation request in the current RSU communication coverage area, The vehicle can only wait until the time for this request is exhausted, and then retransmit after timeout. At this time, the total time from the first transmission to completion of the calculation request of this task includes two parts: point-to-point delay and timeout retransmission time. The calculation method is shown in formula 12, where t_u^{e2e} represents the point-to-point delay of the task in the vehicle u.

$$t_{unsuc} = t_u^{e2e} + Timeout \tag{12}$$

4 Simulation Setup and Analysis

4.1 Simulation Setup

The simulation uses Simulated of Urban Mobility (SUMO) as a traffic simulator, and uses OMNet++ to simulate the network. Use the veins framework to combine the two platforms for use. The MAC layer of the framework is based on the 802.11 protocol and is well developed. This study is divided into two scenarios for analysis. In one scenario, adjacent RSUs are not connected, and in the other scenario, adjacent RSUs are connected via optical fiber and can send messages to each other. The distribution of RSUs in the two scenarios meets that the communication coverage areas of adjacent RSUs do not overlap and are tangent. The important parameters of the simulation settings are listed in Table 1.

Table 1. Simulation setting.

Settings	Values
Max communication distance(m)	500
Number of vehicles	20/40/80
Speed of vehicles(km/h)	60/80/100/120
Number of edge servers	3
Number of cloud servers	1
$t^{e2e}(s)$	0.04
Simulation time(s)	180

4.2 Performance Analysis

We simulated the two scenarios several times and analyzed the simulation results. When 20 cars are added to the simulation scene and there is no collaboration between RSUs, the simulation results are shown in Fig. 5. When the vehicle has a stable connection with the RSU, the message delay is about 0.02 s. When the vehicle unloads the calculation task to the RSU and leaves the communication coverage of the RSU before receiving the calculation result, the task delay will increase to between 0.11 s and 0.12 s. This is because the vehicle does not get the result of the task calculation at this time, causing the vehicle to wait for a timeout and retransmission time before sending the calculation request for this task again.

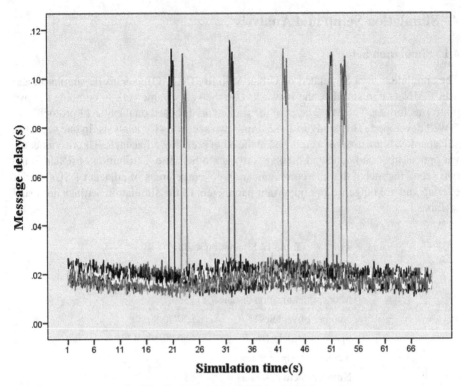

Fig. 5. Task delay of different vehicles

When 20 vehicles are added to the simulation scene and the adjacent RSUs collaborate, the simulation result is shown in Fig. 6.

It can be seen from the result that the message delay is about 0.02 s when the vehicle has a stable connection with the RSU. When a vehicle arrives from an RSU coverage area to an adjacent RSU coverage area and sends a calculation request, the task delay is between 0.09 s and 0.1 s. Due to the cooperation between RSUs, the vehicle can receive the task calculation result when it sends the task calculation request for the first time, and the vehicle does not need to resend the task calculation request.

Figure 7 shows the number of messages that need to be forwarded by adjacent RSUs when different numbers of vehicles are at different speeds during the effective simulation time of 100 s. It can be seen that when the vehicle speed is from 80 km/h to 100 km/h, there is a significant increase in the number of messages. Therefore, adjacent RSUs need to be deployed to connect to each other in the road sections where the speed is about 100 km/h and greater than 100 km/h, so that the RSUs can cooperate to reduce the delay of task processing.

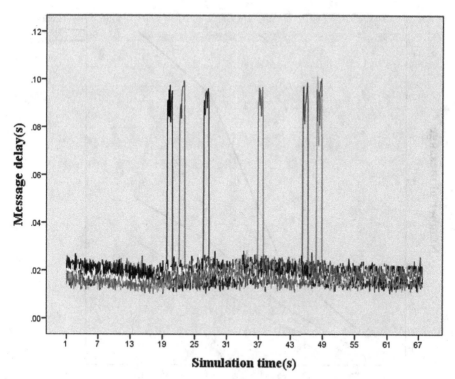

Fig. 6. Task delay of different vehicles

In these two scenarios, the article mainly compares the average delay of task completion when the vehicle leaves the communication coverage area of the current RSU to reach the communication coverage area of the next RSU. Because the side vehicle node cannot receive the calculation result of the task, the task completion time increases sharply. In the second scenario, using the cooperation between adjacent RSUs, the vehicle can receive the calculation result of the previous RSU in the next RSU, which can effectively reduce the task delay caused by retransmission timeout. Therefore, when the RSU is not cooperating, the task completion time includes two parts: retransmission timeout delay and message point-to-point delay. Figure 8 shows the comparison of the average time to complete tasks for 40 vehicles at different speeds.

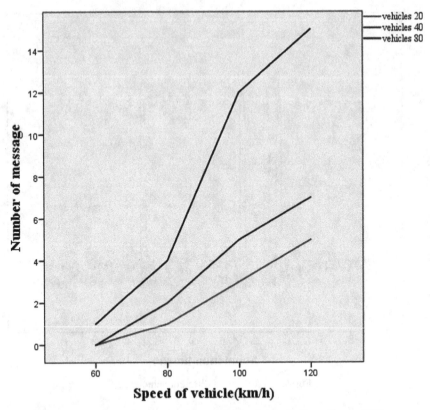

Fig. 7. Number of messages that need to be forwarded by adjacent RSUs

It can be seen from the simulation results that cooperative communication between adjacent RSUs can reduce the time for task completion. The algorithm for calculating the remaining time of the vehicle in the current RSU communication range based on the distance between the vehicle and the RSU is effective. Compared with the scenario where the RSU does not work together, the communication model and algorithm proposed in this paper reduce the delay by 25%, which can greatly improve the performance of vehicle applications.

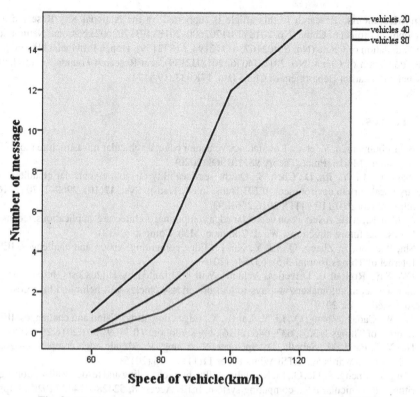

Fig. 8. Number of messages that need to be forwarded by adjacent RSUs

5 Conclusion

In order to reduce the delay of the task request of the moving vehicle during the RSU switching process, this paper proposes a cooperative communication method between adjacent RSUs. In this communication method, an algorithm for calculating the remaining travel time of the vehicle in the RSU communication range based on the distance between the vehicle and the RSU is proposed. The algorithm screens out messages that need to be transmitted between adjacent RSUs, the purpose is to reduce the number of data transmissions between adjacent RSUs, thereby reducing the network load between RSUs and speeding up data transmission. In order to test the performance of the communication model and algorithm, this article uses SUMO to establish a traffic scene, uses OMNet++ for simulation experiments, and establishes two different scenes for comparison. The simulation result proves the feasibility and effectiveness of the cooperative communication model between adjacent RSUs and the vehicle remaining driving time algorithm in the current RSU.

Acknowledgments. Research in this article is supported by the National Key Research and Development Project of China (No. 2018YFB1702600, 2018YFB1702602),National Natural Science Foundation of China (No. 61402167, 61772193, 61872139), Hunan Provincial Natural Science Foundation of China (No. 2017JJ4036, 2018JJ2139), and Research Foundation of Hunan Provincial Education Department of China (No. 17K033, 19A174).

References

1. Balasubramanian, V., et al.: Low-latency vehicular edge: a vehicular infrastructure model for 5G. Simul. Model. Pract. Theory **98**, 101968 (2020)
2. Wang, C., Li, Y., Jin, D., Chen, S.: On the serviceability of mobile vehicular cloudlets in a large-scale urban environment. IEEE Trans. Intell. Transp. Syst. **17**(10), 2960–2970 (2016). https://doi.org/10.1109/TITS.2016.2561293
3. Salman, R., et al.: A survey on vehicular edge computing: architecture, applications, technical issues, and future directions. Wirel. Commun. Mob. Comput. (2019)
4. Shi, W., Cao, J., Zhang, Q., Li, Y., Xu, L.: Edge computing: vision and challenges. IEEE Internet of Things Journal **3**(5), 637646 (2016)
5. VW Says Rollout of Driverless Vehicles Will be Limited by High Costs. https://europe. autonews.com/automakers/vw-says-rollout-driverless-vehicles-will-belimited-high-costs. Accessed 5 Mar 2019
6. Shi, W., Cao, J., Zhang, Q., Li, Y., Lanyu, X.: Edge computing: vision and challenges. IEEE Internet of Things J. **3**(5), 637–646 (2016). https://doi.org/10.1109/JIOT.2016.2579198
7. Hu, Y.C., Patel, M., Sabella, D., Sprecher, N., Young, V.: Mobile edge computing—Akey technology towards 5G. ETSI White Paper **11**(11), 1–16 (2015)
8. Wang, Z., Zheng, S., Ge, Q., Li, K.: Online offloading scheduling and resource allocation algorithms for vehicular edge computing system. IEEE Access **8**, 52428–52442 (2020). https://doi.org/10.1109/ACCESS.2020.2981045
9. Chen, X.: Decentralized computation offloading game for mobile cloud computing. IEEE Trans. Parallel Distrib. Syst. **26**(4), 974983 https://doi.org/10.1109/TPDS.2014.2316834
10. Zheng, J., Cai, Y., Yuan, W., Shen, X.: Dynamic computation offloading for mobile cloud computing: a stochastic game-theoretic approach. IEEE Trans. Mob. Comput. **18**(4), 771–786 (2019). https://doi.org/10.1109/TMC.2018.2847337
11. Wang, Y., Sheng, M., Wang, X., Wang, L., Li, J.: Mobile-Edge Computing: Partial Computation Offloading Using Dynamic Voltage Scaling. IEEE Trans. Commun. **64**(10), 4268–4282 (2016). https://doi.org/10.1109/TCOMM.2016.2599530
12. Dinh, T.Q., Tang, J., La, Q.D., Quek, T.Q.: Offloading in mobile edge computing: Task allocation and computational frequency scaling. IEEE Trans. Commun. **65**(8), 3571–3584 (2017). https://doi.org/10.1109/TCOMM.2017.2699660
13. Zhang, K., Mao, Y., Leng, S., He, Y., Yan, Z.H.A.N.G.: Mobile-edge computing for vehicular networks: a promising network paradigm with predictive off-loading. IEEE Vehicular Technol. Mag. **12**(2), 36–44 (2017). https://doi.org/10.1109/MVT.2017.2668838
14. Wan, S., Li, X., Xue, Y., Lin, W., Xiaolong, X.: Efficient computation offloading for Internet of Vehicles in edge computing-assisted 5G networks. J. Supercomput. **76**(4), 2518–2547 (2020). https://doi.org/10.1007/s11227-019-03011-4
15. Wang, Y., et al.: A game-based computation offloading method in vehicular multiaccess edge computing networks. IEEE Internet Things J. **7**(6), 4987–4996 (2020). https://doi.org/10. 1109/JIOT.2020.2972061
16. Zhao, J., Li, Q., Gong, Y., Zhang, K.: Computation Offloading and Resource Allocation For Cloud Assisted Mobile Edge Computing in Vehicular Networks. IEEE Trans. Veh. Technol. **68**(8), 7944–7956 (2019). https://doi.org/10.1109/TVT.2019.2917890

Author Index

Printed in the United States
by Baker & Taylor Publisher Services